Contents

Preface

Approximately two years ago, several of my larger industrial clients asked me to train their electrical-maintenance personnel in the fundamentals of industrial electronic automation. Before starting a formal industrial training course, I held a mini-seminar to determine how many other local industries would also be interested in such a program. The response was overwhelming. To accommodate the number of participants, two classes had to be formed. The total number of participants exceeded the expected number of participants by about 15 percent. Of those attending the mini-seminars, about 80 percent said they would enroll in a comprehensive industrial-electronics training program.

In designing this training program, my first objective was to find an appropriate textbook that provided a good overview of modern industrial electronic automation. This was much more difficult than I anticipated. Although many specialized textbooks were available, I could not find a suitable primer that encompassed the entire field. Therefore, I began writing weekly handouts to supplement the lecture sessions of the training program. Local participating industries aided me greatly in this effort by providing feedback on industry training needs and on the day-to-day progress of their involved employees. After minor revisions, the weekly handouts were compiled, forming this book.

I would like to thank E.I. DuPont de Nemours & Co., Waldorf Corporation (formerly Champion International Corporation, Folding Cartons Division), and Custom-Pak Incorporated for their cooperative efforts during the training program.

In addition, a special thanks goes out to Christi Koopman for her invaluable literary assistance and marketing research.

Introduction

We all learn by doing. This is the reason why most employers require a minimum number of years of experience for potential employees. Practical experience is essential for long-term success. Practical experience is what this book is all about.

Most electrical engineers forget the majority of high-level, seldom-used formulas and equations they spent many nights struggling to learn in college. This is because they replace complicated formulas and equations with practical rules-of-thumb. They develop logical design and troubleshooting methods through experience.

After 14 years of industrial engineering, as well as training of industrial-engineering and maintenance personnel, I have collected all of my shortcuts, methods, rules-of-thumb, and aids to understanding in this book. The book's objective is to provide the reader with as much industrial engineering data and experience as possible in one easy-to-understand manual.

This book is designed to provide readers with the basic knowledge needed for proficiency in this highly specialized field. I do not assume that the reader is or desires to be an electrical engineer. My intention is to inform the reader, regardless of background, in three basic areas:

1. The workings of industrial electronics.
2. Industrial-electronics applications.
3. Industrial-electronics repair and servicing.

These topics are taught in a down-to-earth manner that is easily understood by electrical maintenance personnel, management personnel, or virtually anyone with the desire to learn. The book begins with the basics of electronics and progresses toward a complete systems understanding.

Most importantly, the information and troubleshooting methods can be implemented immediately. Although there is no substitute for hands-on experience, this book provides the reader with concentrated information that would otherwise take years of trial-and-error experience to obtain.

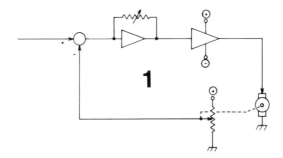

The Competitive World of Industrial Electronic Automation

INDUSTRIAL ELECTRONIC AUTOMATION IS THE USE OF ELEC-tronics to enable industrial manufacturing machinery to operate without constant supervision. The most common type of industrial electronic automation monitors a product being manufactured and automatically corrects any errors detected in the product. For example, in the manufacture of a plastic sheet requiring a uniform thickness, industrial electronic automation can be used to measure the deviation from the desired thickness and automatically correct any excessive errors detected. This automatic correction would be performed without human intervention. Also, the measurement and correction would be performed much more quickly and with a higher degree of precision than would be humanly possible.

In some cases, it is necessary to automatically monitor and control the manufacturing machine's performance rather than the quality of a product. For example, you may wish to accurately maintain the speed of a motor to provide optimum performance of a manufacturing machine. A modern electronic motor control can sense the rotational speed of a motor and automatically adjust this speed to the desired speed.

The previous examples represent a very small sampling of the many types of industrial electronic automation currently being used in modern manufacturing. But why is industrial electronic automation so important to society? Is it a threat or benefit to human labor?

HIGH-TECH AUTOMATION: FRIEND OR FOE?

For the first time in history, humans are being intimidated by intelligent machines. These machines fit into the category of high-technology automation, and their incorporation has created a subtle shuffle of manpower utilization throughout the world. This has caused people to fear losing their jobs to machines.

Industrial automation began during the Industrial Revolution, when Henry Ford pioneered the concept of mass production. The greatest advantage of mass production was that it allowed a person to become proficient at one facet of a complete manufacturing process. This greatly improved the efficiency of the process because the assembly line workers did not have to be multitalented. Unfortunately, it did not eliminate the manufacturing errors originating from human error caused by fatigue and boredom. Human speed was another limitation.

Society is undergoing a second "industrial revolution." Most human speed and error limitations can be eliminated in today's high-tech industrial processes. High-speed computers, process controllers, robots, and other electronic machines eliminate the need to contend with monotonous, boring, and dangerous operations.

Does this mean that human labor is becoming obsolete? Definitely not! The need for humans in industry is more prevalent than ever. This is because the need for the human mind still exists. The function of automation is to free humans from mundane tasks and improve upon the efficiency and utilization of time. Thus it enables humans to concentrate on more important issues. Automation also provides a degree of speed and precision that would be impossible with manual labor.

Actually, the lack of automation is a much greater threat to the labor force in today's world than the loss of a small number of unskilled labor positions. The reason for this is twofold. First, the incorporation of a significant amount of electronic automation within an industry increases productivity, creating openings for new labor positions. Secondly, the lack of automation destroys a company's ability to compete with more progressive companies. Obviously, if a company cannot compete, it eventually goes out of business, leaving all of its employees out of work.

The following account of an actual installation illustrates this point. A multimillion-dollar continuous-sheet-process control system was purchased by a small division (consisting of approximately 80 employees) of a large corporation. (The name of the corporation shall remain confidential.) Before the system was installed, the small division had shown a loss and had been in danger of being closed. Three months after the installation had been completed, the division began showing its first profit in almost two years. Although the process-control system had replaced the duties of six employees, the increased productivity had simultaneously created additional manpower requirements in the shipping, receiving, and purchasing departments. This resulted in the creation of 10 new positions, bringing the total number of employees to 84.

Unfortunately, the implementation of modern automation equipment does not always create as many jobs as it eliminates. A price must be paid for survival. But the number of jobs saved by automation greatly exceeds the number of jobs replaced by it in the long run.

Consider the technological competition between the U.S. and Japan from 1960 to 1980. Although the United States made all of the major advancements in the equipment and fundamentals of the automation field, it was much slower than Japan in implementing these breakthroughs. In 1960, the U.S. produced approximately 8,000,000 motor vehicles, which accounted for about 48 percent of the world marketplace. At the same time, Japan was capturing about 3 percent of the world's auto sales. In 1980, the U.S. produced about the same number of motor vehicles, but only captured 20 percent of the market while Japan's share of the market had increased to 28 percent. From 1981 to 1982, the U.S. auto industry laid off over 40,000 management personnel and 250,000 union workers due to loss of sales.

In 1985, the United States had the largest gross national product (GNP) of any country in the world. During the past five years, this GNP has shown an annual increase of 2 to 3 percent per year. In contrast, Japan is currently the most technologically advanced manufacturing country in the world. (Japan possesses more than one-half of the world's industrial robots.) During the past five years, Japan's GNP has shown an extraordinary growth of 6 to 10 percent per year. If these statistics remain the same during the next few years, Japan's GNP will equal the GNP of the United States sometime between 1987 and 1988! Germany, Italy, and France are presently enjoying an increase in their annual GNP's of approximately 4 to 5 percent per year.

This is not to imply that a lack of automation is the only reason for the dramatic loss of sales by U.S. industries, but it is certainly at the top of the list. While high-ranking government politicians argue about the complex economic situations creating the decline of U.S. prominence in the world marketplace, U.S. industries are investing the largest percentage of total capital expenditures in history toward new automation. (From 1981 to 1982, the U.S. automotive industry spent more money on automation than has been spent on the entire U.S. space-shuttle program.)

Simply stated, automation is here to stay. American industries are in a desperate struggle to catch up with the automation technologies presently being used by their foreign competitors (primarily Japan, France, Germany, and Italy).

Whether the main purpose of automation is reduced cost, improved documentation, or test repeatability, the MP2903 can quickly automate unique applications (courtesy of Tektronix, Inc.).

PRESENT AUTOMATION IN THE UNITED STATES

Technologically speaking, depending on the industry, Japan is currently 5 to 10 years ahead of the United States. However, American industries are attempting to get back into the competitive world marketplace, both monetarily and technologically. This is evident by the $13 billion being spent annually in the U.S. on automation equipment and services.

The United States is presently in a transitional stage of automation development. Because U.S. industries were slow in realizing the competitive edge of their foreign competitors, they are catching up as efficiently as possible. A large portion of this catching-up process involves upgrading existing manufacturing equipment with intelligent control systems. For this reason, it is common to see an antiquated piece of machinery being controlled by a state-of-the-art computer. Depending on the application, a computer (or similar microprocessor-based control system) can compensate for many of the shortcomings associated with machinery age or wear.

Another promising step is a change of attitude on the part of American industries and their employees. American industries have had to face the harsh reality that they no longer control the world

Continuing advancements in integrated circuit manufacturing technology have elevated the automation field to a highly complex technology (courtesy of Fairchild Semiconductor Corporation).

marketplace; thus, they are taking steps to fit into it. In addition, many industries are incorporating modern employee interaction programs to improve company-employee relations. These interaction programs concentrate on employee health, safety, and mental attitude to improve employee efficiency and quality of work.

Fortunately, U.S. industries have realized that they must modernize their way of thinking along with their manufacturing equipment to effectively compete in today's world marketplace.

THE FUTURE OF AUTOMATION

Without question, the future promises to be a golden age of automation. Although you can expect to see many innovations in the consumer marketplace, the most dramatic advances will occur in industry. Manufacturers who take advantage of these advances will prosper, while the rest will be left behind.

Electronics experts agree that future advancements in automation are not going to be related to mechanical movements.

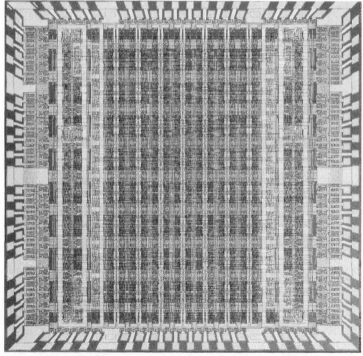

Microscopic photograph of the complexity of an integrated circuit (courtesy of Fairchild Semiconductor Corporation).

Rather, present trends indicate that future automation will progress toward mechanically simple, electronically complex machines. There are several reasons for this. First, it is cheaper to manufacture smart (electronically complex) machines than it is to manufacture mechanically complex machines. Secondly, smart machines are more versatile. The overall operation of a smart machine can be changed by simply loading in a new program.

The computer field will continue its rapid growth. Advancements in computer hardware, along with parallel advancements in computer software (programs), will increase computer utilization to a level of sophistication that would have been almost unimaginable just a few short years ago.

The field of industrial robotics promises to be one of the most exciting and rapidly growing fields of the future. Not only will industrial robots gain greater intelligence and sensory abilities, but they will also become mobile (able to move from one work station to another).

YOUR FUTURE AND AUTOMATION

So how do you, the reader, fit into this future picture? In the

Distributed manufacturing control system controls industrial equipment and integrates the equipment into information management systems (courtesy of Honeywell Process Control Division, Fort Washington, PA 19034).

future, industrial electronic automation will affect you in many ways. For example, unskilled jobs in hazardous or highly uncomfortable environments will likely be performed by automated machinery. This will improve overall working conditions and reduce accidents. You may find yourself monitoring an automated machine instead of doing manual work.

Your on-the-job training opportunities will probably increase. Companies will need trained personnel to operate and/or maintain hi-tech equipment.

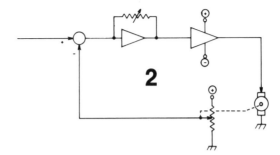

2

Characteristics of Electricity

AT SOME POINT IN YOUR LIFE, YOU HAVE PROBABLY FELT electricity. If this experience involved a mishap with 120 Vac (common household voltage) or higher, you already know that a physical force is associated with electricity.

You constantly see the effects of electricity when you use electrical devices such as motors, electric heaters, television sets, and dishwashers. But even with the almost constant usage of electricity, it often possesses an aura of mystery because it cannot be seen. The best you can do is explain its characteristics by comparing these characteristics with something that you can see and with which you are familiar. See Fig. 2-1.

Examine the simple water-flow system shown in Fig. 2-1. This hypothetical system consists of a water pump, a valve to turn the water flow on and off, a pipe to carry the water to the valve, and pipe from the valve to return the water to the pump. Assuming the pump runs continuously and the valve can only be fully open or fully closed, there can be two possible conditions within this system; the valve is closed, restricting all flow, or the valve is open, allowing water to flow throughout the system.

If the valve is closed, no water will flow through the pipe, but water pressure (from the running pump) will still exist. The valve resists the flow of water.

With the valve fully open, water flow will exist with virtually no resistance from the valve. The rate of this flow will be directly

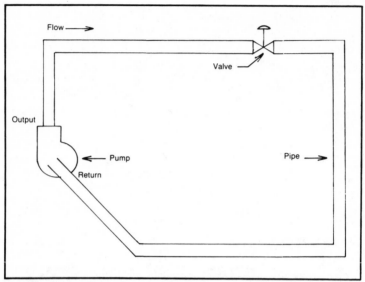

Fig. 2-1. Fluid analogy of an electric circuit.

proportional to the pump capacity and the amount of resistance to the water flow. The only restriction to infinite water flow in an open valve is the size (diameter) of the pipe and the maximum pressure the pump is capable of producing. See Fig. 2-2.

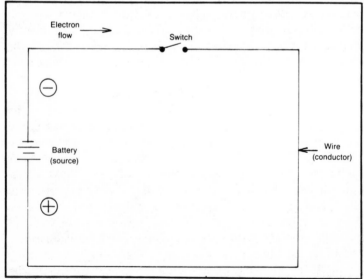

Fig. 2-2. Basic electrical circuit.

The electrical circuit shown in Fig. 2-2 is very much like the water system shown in Fig. 2-1. The battery is analogous to the pump. It provides an electrical pressure that produces an electrical flow of current through the wire. The electrical pressure is called *voltage*. The electrical flow is called *current*.

If the switch is open, it will not allow current to flow because no electrical path will exist. In other words, since a conductive path does not exist from one end of the circuit to the other, there will be no electrical continuity.

If the switch is closed, it will connect one end of the wire to the other, and current will begin to flow. The rate, or amount, of current flow will depend on the battery voltage and the size (diameter) of the wire. If the wire diameter is reduced, the wire's electrical *resistance*, or opposition to current, will increase and, therefore, the current flow will decrease.

Thus, in any electrical circuit, there will always be three variables to consider: voltage (electrical pressure), current (electrical flow), and resistance (the opposition to current flow). These variables can now be examined in greater detail.

VOLTAGE

It is important to keep in mind that voltage does not move; rather, it is applied. This statement may seem abstract at first, but consider the following comparison. Imagine a two-inch diameter tube one mile long, filled with ping-pong balls. If you should insert one more ping-pong ball in one end, one ping-pong ball will be forced out of the other end. As you insert the additional ping-pong ball, the pressure causing all of the ping-pong balls to move within the tube is applied at the same time.

The movement of the ping-pong balls is analogous to the current flow. The pressure causing them to move is analogous to the voltage. The pressure did not move; it was applied. Only the ping-pong balls moved. The same is true of voltage and current. Voltage does not move; only current does.

Voltage is often referred to as *electrical potential*, and its proper name is *electromotive force*. It is measured in units called volts. Its electrical symbol, as used in formulas and expressions, is E.

The level, or *amplitude*, of voltage is usually defined in respect to some common point. For example, the battery in most automobiles provides approximately 12 volts of electrical pressure. The negative side of the battery is normally connected to the main body of the automobile, causing all of the metal parts connecting to the body and frame to become the common point of reference. When measuring the positive terminal of the battery with respect to the body, a positive 12-volt potential will be seen.

This measurement is similar to that of altitude. The point of reference is always sea level. An airplane flying at a 10,000-foot altitude is actually flying at 10,000 feet above sea level. Similarly, voltage is usually expressed as being at a level above (or in excess of) a common reference point.

The terms *positive* or *negative* are used to define the polarity of the voltage. Referring to Fig. 2-1, note how the direction of the water flow is dependent upon the direction, or orientation, of the pump. If the pump had been turned around so that the output and return were switched, the direction of water flow would also have been reversed.

Another way of looking at this condition is to consider the output side of the pump as having a positive pressure associated with it while the return has a negative pressure (vacuum) associated with it. The water will be sucked toward the vacuum and be pushed away from the output.

In simple terms, the polarity of an electrical potential determines the direction in which current will flow. But it is important to note that the direction of the current flow is opposite from what you may expect. In an electrical circuit, current will always flow from a negative potential to a positive potential. This is referred to as an *electron flow*. The direction of electron flow is shown in Fig. 2-2.

AC AND DC VOLTAGES

The periodic reversal of current flow is called *alternating current* (ac). In ac-powered circuits, the polarity of the voltage changes perpetually at a specified rate, or frequency. Since the polarity of the voltage is what determines the direction of the current flow, the current flow changes directions at the rate at which the voltage polarity changes. The symbol for an ac voltage is shown in Fig. 2-3.

The frequency of current alternations is measured in units called *hertz* (Hz). A synonym for hertz is *cycles per second* (cps). The terms *hertz* and cycles per second both define how many times the current reverses direction in a one-second time period. Common power of the average home in the United States has been stan-

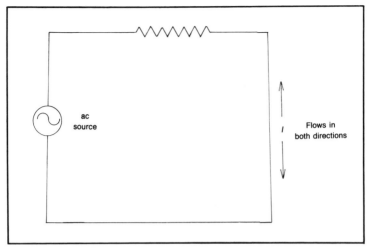

Fig. 2-3. Basic ac circuit illustrating current flow in both directions.

dardized at 60 Hz. This means the voltage polarity and current flow reverse direction 60 times every second. See Fig. 2-3.

In contrast to ac voltages, a *direct-current* (dc) voltage never changes its polarity. For this reason, the current flow likewise never changes its direction. The battery shown in Fig. 2-2 produces a dc voltage, as do all batteries.

CURRENT

Of the three previously mentioned variables, current is the only one that moves, or flows. Its flow rate is measured in units called *amperes*, or *amps*. Its electrical symbol as used in formulas and expressions is I. As illustrated in Fig. 2-4, current can only flow in a *closed circuit*—that is, a circuit that provides a continuous conductive path from the negative potential to the positive potential. If there is a break in this continuous path, such as an open switch, current cannot flow, and the circuit is said to be *open*. See Fig. 2-5.

RESISTANCE

Resistance is the opposition to current flow. The switch in the open circuit shown in Fig. 2-5, actually represents an infinite resistance. In other words, the switch resistance is so high that it does not allow current to flow. The closed switch shown in Fig. 2-4 is an example of the opposite extreme. It has virtually no resistance to current flow and, therefore, no effect upon current flow within the circuit.

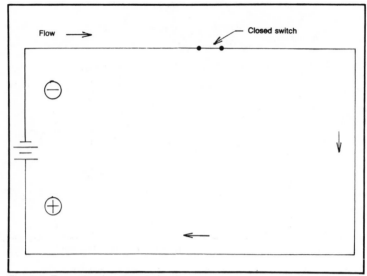

Fig. 2-4. Example of a closed circuit.

The circuit illustrated in Fig. 2-6 shows the symbol for a *resistor* and the current (*I*) flow through the circuit. A resistor will present some resistance to current flow that falls between the two extremes presented by an open or closed switch. This resistance will normally be much higher than the resistance of the wire used to connect the circuit together. Therefore, under most circum-

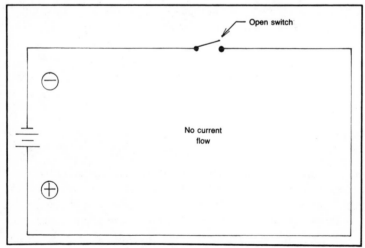

Fig. 2-5. Example of an open circuit.

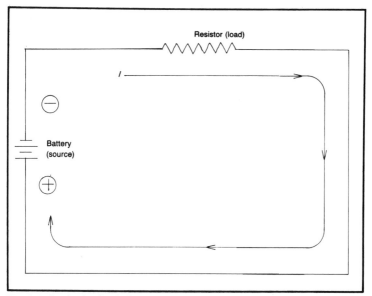

Fig. 2-6. Basic dc circuit illustrating the direction of current flow.

stances, this wire resistance is considered to be negligible. Resistance is measured in units called *ohms*. Its electrical symbol is *R*.

CONDUCTANCE

Sometimes it is more convenient to consider the amount of current allowed to flow rather than the amount of current opposed. In these cases, the term *conductance* is used. Conductance is simply the reciprocal of resistance. The unit of conductance is the *mho* (ohm spelled backwards), and its electrical symbol is *G*. The following equations show the relationship between resistance and conductance (for example, 2 ohms of resistance would equal 0.5 mhos of conductance and vice versa):

$$\frac{1}{G} = R$$

in which the reciprocal of conductance (*G*) is resistance (*R*);

$$\frac{1}{R} = G$$

in which the reciprocal of resistance (*R*) is conductance (*G*).

In recent years, the accepted term of conductance, mho, and its associated symbol, *G*, have been replaced with the term *siemen* and its associated symbol (*S*). These two terms mean exactly the same thing. Both are the reciprocal of resistance. However, throughout the remainder of this book, I will continue to use the older term mho, which is still the most commonly used term.

POWER

The amount of energy dissipated (used) in a circuit is called *power*. Power is measured in units called *watts*. Its electrical symbol is *W*. Electrical power is dissipated as heat.

In circuits such as the one shown in Fig. 2-6, the resistance is often referred to as the *load*. The battery is called the *source*. This is simply a means of explaining the origin and destination of the electrical power. For example, in Fig. 2-6, the electrical power comes from the battery. Therefore, the battery is the source of the power. All of this power is going into the resistor. Thus, the resistor is the load.

You have briefly examined the basic units of measurement for resistance, voltage, current, and power. These units are the ohm, volt, amp, and watt respectively. In the electrical and electronic fields, you will have to work with extremely large and extremely small quantities pertaining to these basic units. To simplify dealing with such quantities, common metric prefixes are used to aid in describing them. These prefixes are listed below. You should memorize these before proceeding to the next chapter.

$$
\begin{aligned}
&\textit{pico} \ (\text{symbol } p) = 1/1,000,000,000,000 \\
&\textit{nano} \ (\text{symbol } n) = 1/1,000,000,000 \\
&\textit{micro} \ (\text{symbol } \mu) = 1/1,000,000 \\
&\textit{milli} \ (\text{symbol } m) = 1/1,000 \\
&\textit{kilo} \ (\text{symbol } k) = 1,000 \\
&\textit{mega} \ (\text{symbol } M) = 1,000,000
\end{aligned}
$$

EXAMPLES:

1 picovolt	=	.000000000001 volt	(abbrev. 1 pV)
1 nanoamp	=	.000000001 amp	(abbrev. 1 nA)
1 microvolt	=	.000001 volt	(abbrev. 1 μV)
1 milliamp	=	.001 amp	(abbrev. 1 mA)
1 kilovolt	=	1000 volts	(abbrev. 1 kV)
1 megawatt	=	1,000,000 watts	(abbrev. 1 MW)

For clarity, some explanations are needed. The Greek symbol μ (pronounced "mu") is used to avoid confusion when abbreviating *micro* and *milli*. In many cases, abbreviations are totally eliminated on schematics and written material. Also, note that prefix abbreviations for quantities greater than 1 are capitalized, and prefix abbreviations for quantities less than one are written in lowercase letters. For example, the abbreviation for *kilo* is K, and the abbreviation for *milli* is m.

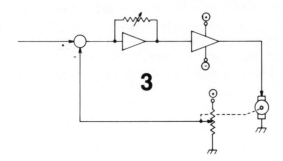

3

Laws of Electricity

AS WITH ALL OTHER PHYSICAL FORCES, PHYSICAL LAWS GOV-
ern electrical energy. Highly complex electrical-engineering
projects may require the use of several types of high-level
mathematics. The electrical-design engineer must be well acquainted with physics, geometry, calculus, and algebra.

However, if you do not possess a high degree of proficiency
with high-level mathematics, this does not mean you must aban-
don your present goal of becoming proficient in the knowledge and
techniques of industrial automation electronics. In fact, many people who have achieved success in this field have done so without
the knowledge of high-level math. Unfortunately, this does not
mean that you can ignore math altogether. Math is necessary for
establishing the definite physical relationships between voltage, current, resistance, and power.

OHM'S LAW

The most basic form of defining electrical relationships is called
Ohm's law. In the electrical/electronic fields, Ohm's law is a basic
tool for comprehending electrical circuits and analyzing problems.
Therefore, it is important to memorize and become familiar with
the proper use of Ohm's law just as a carpenter must learn how
to properly use a saw. Ohm's law is shown in Equation 3-1.

$$E = IR \qquad \textbf{Equation 3-1}$$

in which voltage (*E*) equals the current value (*I*) multiplied by the resistance value (*R*).

If you are not familiar with algebra, following is an explanation of the above equation. Whenever two or more symbols are placed side by side in an equation, you must multiply these values. For example, in the equation $E = IR$, *I* and *R* are side by side; they do not have a mathematical symbol between them. Therefore, they must be multiplied. If the value of *I* is 2 and the value of *R* is 3, the answer would be 6 (2 × 3 = 6).

When actual number values are substituted in place of the electrical symbols, these numbers will be placed in parentheses to separate them. If there is no mathematical symbol between them, it is assumed they should be multiplied together (for example, (2)(3) = 6).

SERIES AND PARALLEL CIRCUITS

To fully appreciate Ohm's law, see Fig. 3-1. The circuit shown here illustrates a condition in which 1 amp of current flows through a resistance of 10 ohms. By substituting the electrical symbols in the equation with the actual values, the result is:

$$E = I\,R$$
$$E = (1)\,(10)$$
$$E = 10 \text{ volts}$$

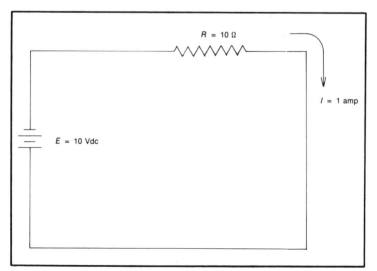

Fig. 3-1. Basic dc circuit illustrating Ohm's law.

NOTE: In Fig. 3-1, the Greek omega symbol is used to abbreviate ohms.

Figure 3-2 shows another example. Here, the equation can be applied as follows:

$$E = I R$$
$$E = (0.5)\ (6.8)$$
$$E = 3.4 \text{ volts}$$

NOTE: In Figs. 3-1 and 3-2, a battery was used as the source. Therefore, the voltage will be a dc voltage. This is usually abbreviated Vdc.

Another important rule of algebra is that you may do whatever you want to one side of an equation as long as you do the same to the other side of the equation. This keeps the equation equal, or balanced, which is essential.

Suppose you knew the voltage and resistance, but not the current, in Fig. 3-1. To find the current, you would have to isolate I on one side of the equation. This can be accomplished by first dividing both sides of the equation by R as follows:

$$\frac{E}{R} = \frac{IR}{R}$$

The two R's on the right side of the equation cancel each other out. Therefore:

$$\frac{E}{R} = I \quad \text{or} \quad I = \frac{E}{R}$$

Now you can solve for the current (I) value.

$$I = \frac{E}{R}$$
$$I = \frac{10 \text{ volts}}{10 \text{ ohms}}$$
$$I = 1 \text{ amp}$$

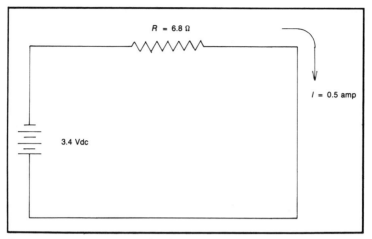

Fig. 3-2. Example illustrating Ohm's law.

Similarly, you could rearrange the equation to solve for resistance if you knew the current and voltage. Note that each side of the equation must be divided by I.

$$\frac{E}{I} = \frac{IR}{I}$$

The two I's on the right side of the equation cancel out each other leaving:

$$\frac{E}{I} = R \quad \text{or} \quad R = \frac{E}{I}$$

Now you can solve for the resistance (R) value.

$$R = \frac{E}{I}$$

$$R = \frac{10 \text{ volts}}{1 \text{ amp}}$$

$$R = 10 \text{ ohms}$$

By understanding the above equations, you can now fully understand what is meant by a proportional relationship between voltage, current, and resistance. If one of the values is changed, at least one of the other values must also change. For example, if you increase the source voltage, the current must increase. If you increase the resistance, the current must decrease. For practice, try substituting other values for the current, voltage, and resistance, and notice how this relationship works.

Figure 3-3 is an example of a *parallel circuit*. A parallel circuit is a circuit in which one resistance is electrically placed across another resistance. The basic concept of a parallel circuit can be compared to the water-flow system shown in Fig. 3-4. In this system, the pump applies the pressure to move the water through the two valves. The valves are placed in parallel to each other and are adjusted to allow 1 gallon per minute (GPM) of water flow through them. The main water line leading to and from the pump must carry the total flow, which is the sum of the flows going through each valve. In this example, there are two valves, and the flow through each valve is 1 GPM. Therefore, the total flow must be 2 GPM. It is important to note that the same water pressure is applied to each valve.

Refer to the circuit shown in Fig. 3-3. In principle, the same condition exists as in the water-flow system of Fig. 3-4. The battery supplies the electrical pressure (voltage) that causes current

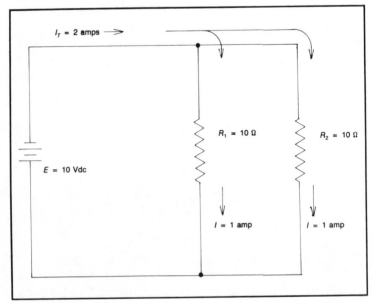

Fig. 3-3. Simple parallel dc circuit.

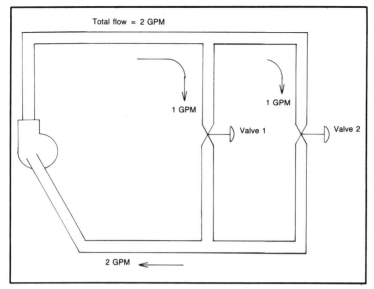

Fig. 3-4. Fluid analogy of a parallel electric circuit.

to flow. Through each resistance leg of this circuit, there will be 1 amp of current flow. Therefore, the total current flow must be the sum of the currents flowing through the individual legs, or 2 amps. Also, as in the case of the water-flow example, the same voltage (electrical pressure) is applied to each resistance.

Using Ohm's law, the relationship illustrated in Fig. 3-3 can be proven to be correct. As previously stated, the voltage across R_1 is equal to the voltage across R_2. In this case, this voltage is the source voltage of 10 Vdc. Therefore, the current flow through R_1 is:

$$I = \frac{E}{R} = \frac{10 \text{ volts}}{10 \text{ ohms}} = 1 \text{ amp}$$

Since the voltage and resistance values are the same for R_2, the current flow through R_2 must also be 1 amp. Therefore:

$$I_{total} = 1 \text{ amp} + 1 \text{ amp} = 2 \text{ amps}.$$

When two (or more) resistances are put in parallel, their combined effect within the circuit must be considered. In other words, it is necessary to consider their equivalent resistive effect as if they were combined into one resistor. In a parallel circuit, this combined resistance will always be less than either of the resistances alone. Ohm's law can again be used to calculate this combined (or equivalent) resistance for Fig. 3-3.

NOTE: In this case, the equivalent resistance will be the total circuit resistance. Therefore, when performing the equation, the total values for voltage and current must also be used. In Fig. 3-3, the total voltage is the same as the source voltage.

$$R_{total} = \frac{E_{total}}{I_{total}} = \frac{10 \text{ volts}}{2 \text{ amps}} = 5 \text{ ohms}$$

What has been proven in the above equation is that the equivalent resistive effect of placing two 10-ohm resistors in parallel is actually 5 ohms. Whenever resistors of the same value are placed in parallel, their combined resistance will always equal the value of one of the resistors divided by the number of resistors placed in parallel. In the above example, there were two resistors, and each resistor had a value of 10 ohms. Therefore, the equivalent resistance was equal to 5 ohms (10 divided by 2). If four 10-ohm resistors were placed in parallel, the equivalent resistance would be equal to 10 divided by 4, or 2.5 ohms. This rule will always hold true if the parallel resistances are equal.

NOTE: The equivalent resistance in the parallel circuit will always be less than the resistance in any one branch, or leg, of the circuit.

Unfortunately, most electrical circuits do not contain equal parallel resistances. In the cases where you wish to calculate the equivalent resistive effect (R_{equiv}) of two or more resistances that are not equal you must use either of the following two equations. Equation 3-2 may be used whenever there are only two parallel resistances:

$$R_{equiv} = \frac{(R_1)(R_2)}{(R_1) + (R_2)} \qquad \textbf{Equation 3-2}$$

Equation 3-3 can be used for any number of parallel resistances.

$$R_{equiv} = \cfrac{1}{\cfrac{1}{R_1} + \cfrac{1}{R_2} + \cfrac{1}{R_3}} \qquad \textbf{Equation 3-3}$$

. . . etc.

It is important to understand the principle behind Equation 3-3. You must first find the reciprocal of each resistance value. This reciprocal value will be the conductance value, discussed briefly in Chapter 1. The conductance values are then added together, and their sum is the total conductance value of the parallel circuit. You then calculate the reciprocal of the total conductance, and this answer will be the equivalent resistive effect, or R_{total}.

NOTE: The reciprocal of a given number is found by dividing 1 by that number. For example, the reciprocal of $R = \dfrac{1}{R} = G$ (conductance).

EXAMPLE: Refer to Fig. 3-5. Suppose you want to find the equivalent parallel resistance of 5 ohms and 10 ohms. Equation 3-2 can be used because there are only two resistance values. Following is the solution:

$$R_{equiv} = \frac{(R_1)\,(R_2)}{(R_1) + (R_2)} = \frac{(10)\,(5)}{(10) + (5)} = \frac{50}{15} = 3.33 \text{ ohms}$$

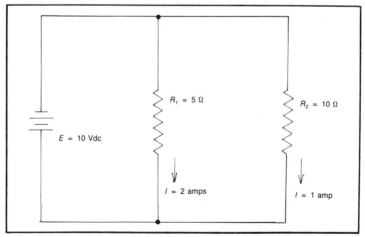

Fig. 3-5. Parallel dc circuit with unequal resistances in each leg.

NOTE: R_{equiv} = 3.33 ohms which is less than either one of the individual parallel resistances.

EXAMPLE: You can also use Equation 3-3 to find the same equivalent resistance of R_1 and R_2 in Fig. 3-5 as follows:

$$R_{equiv} = \cfrac{1}{\cfrac{1}{R_1} + \cfrac{1}{R_2}} = \cfrac{1}{\cfrac{1}{10} + \cfrac{1}{5}} = \frac{1}{(0.1) + (0.2)} = \frac{1}{0.3}$$

$$= 3.33 \text{ ohms}$$

As you can see, either method will work. But remember to perform the calculations in the proper mathematical sequence as shown. In other words, in the above example, you cannot add the values of R_1 and R_2 before you find their reciprocals.

If the resistance values in a parallel network are equal, the current will split evenly through each leg, as shown in Fig. 3-3. But if they are not equal, as shown in Fig. 3-5, the division of the current will be inversely proportional to the resistance value of each individual leg. The term *inversely proportional* simply means that as the resistance value increases, the current flow decreases proportionally.

As stated previously, the applied voltage to a parallel network is equal across all legs of that network. In the circuit shown in Fig. 3-5, the applied voltage to each resistor is the source voltage of 10 Vdc. Since the applied voltage and resistance values are known for each leg, the current flow through each individual leg can be calculated using Ohm's law. Therefore, the current flow through R_1 is:

$$I = \frac{E}{R} = \frac{10 \text{ volts}}{5 \text{ ohms}} = 2 \text{ amps}$$

The current flow through R_2 is:

$$I = \frac{E}{R} = \frac{10 \text{ volts}}{10 \text{ ohms}} = 1 \text{ amp}$$

The total current flow through the circuit of Fig. 3-5 will be the sum of the individual leg currents. Therefore:

$$I_{total} = 2 \text{ amps} + 1 \text{ amp} = 3 \text{ amps}$$

NOTE: You can also calculate the equivalent parallel resistance of the circuit in Fig. 3-5 by using Ohm's law if the total current flow through the parallel resistance network and the applied voltage to it are known. In this case, the total current flow through the parallel network is 3 amps and the applied voltage is the source voltage of 10 Vdc. Therefore:

$$R_{e\,quiv} = \frac{E}{I_{total}} = \frac{10 \text{ volts}}{3 \text{ amps}} = 3.33 \text{ ohms}$$

Based on everything discussed to this point, a general rule governing all parallel circuits can be stated as follows:

In a closed parallel circuit, the applied voltage will be equal across all parallel legs, but the division of the current through each leg will be inversely proportional to the resistance. The equivalent resistance created by the parallel network will always be less than any one resistive leg alone.

The circuit in Fig. 3-6 is an example of a *series circuit*. Notice the difference between this circuit and the circuit shown in Fig. 3-3. In Fig. 3-6, the same current must flow through both resistors. When the same current must flow through all components of

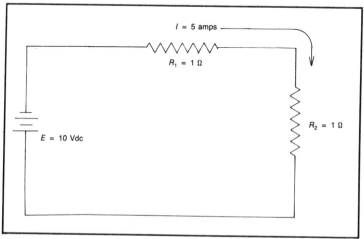

Fig. 3-6. Example of a simple series circuit.

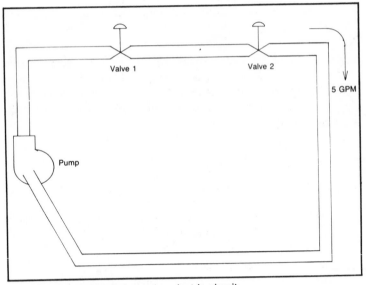

Fig. 3-7. Fluid analogy of a series electric circuit.

a circuit, those components are placed in series with each other. This principle is similar to that of the water-flow system shown in Fig. 3-7. The water must flow through both valves; it cannot branch as it did in Fig. 3-4. There can only be one flow, and this flow must pass through both valves.

The total circuit resistance in Fig. 3-6 is simply the sum of the resistances. Therefore:

$$R_{total} = R_1 + R_2 = 1 \text{ ohm} + 1 \text{ ohm} = 2 \text{ ohms}$$

Because the total circuit resistance is known, you can use Ohm's law to find the total current flowing through the circuit. In a series circuit, this will be the same current flowing through all components in the circuit. Following is the solution:

$$I = \frac{E}{R_{total}} = \frac{10 \text{ volts}}{2 \text{ ohms}} = 5 \text{ amps}$$

In a series circuit, the current is the same through all components, but the division of the voltage is proportional to the resistance. This rule can be proven through the application of Ohm's law. The resistance of R_1 (1 ohm) and the current flowing through it (I_{total} = 5 amps) are known. Therefore, you can now calculate the voltage dropped across it.

$$E (R_1) = I(R_1) R(R_1) = (5) (1) = 5 \text{ volts}$$

NOTE: In the above equation, you are calculating the voltage across R_1. To do this, you must use only those values associated with R_1 to maintain equality in the equation.

You have proven that the voltage dropped across R_1 is 5 volts. Because R_2 has the same resistance value and the same current flowing through it (the current must be the same through all components of a series circuit), the voltage dropped across R_2 must be the same as R_1, or 5 volts. The 5 volts dropped across R_1 and the 5 volts dropped across R_2 equal the source voltage of 10 volts. This demonstrates another very important rule:

The sum of the individual voltage drops across all components in a series circuit must equal the source voltage.

As has been noted previously, the resistors in Fig. 3-6 are equal. However, in most cases, components in a series circuit will not present equal resistances. To calculate the voltage dropped across unequal resistances, first calculate the total circuit resistance by adding the resistive values of all components in the series circuit. Make this R_{total} value the divisor. The dividend is the resistive value of the component for which you will calculate the voltage drop. Perform this division, and then multiply the quotient by the value of the source voltage. The answer is the value of the unknown voltage drop.

To demonstrate the previous method, calculate the voltage dropped across R_1 in Fig. 3-8. You must first calculate R_{total}.

$$R_{total} = R_1 + R_2 = 9 \text{ ohms} + 1 \text{ ohm} = 10 \text{ ohms}$$

Next, make R_{total} the divisor and R_1 the dividend.

$$\frac{9 \text{ ohms}}{10 \text{ ohms}} = 0.9$$

The quotient is then multiplied by the source voltage.

$$(0.9) \ (10 \text{ volts}) = 9 \text{ volts}$$

Therefore, 9 volts will be dropped across R_1. At this point, it should be apparent that 1 volt must be dropped across R_2, because these two voltages must add up to equal our source voltage of 10 Vdc.

In the previous example, you actually calculated the voltage drop across R_1 by forming a ratio (R_1 to R_{total}) and multiplying this ratio by the source voltage. Another method of calculating the voltage dropped across a series resistance is to multiply the circuit current by its resistance value. For example, in Fig. 3-8, the circuit current is 1 amp, and the resistance of R_1 is 9 ohms. Therefore:

$$E = IR = (1) \ (9) = 9 \text{ volts}$$

Series and parallel circuits are often combined to form a *series-parallel circuit*. Figure 3-9 shows an example of a series-parallel circuit. To analyze this circuit, you must begin by finding the equivalent resistance of R_2 and R_3. In this case, the resistances are equal, so you can simply divide either resistance value by 2.

$$R_{equiv} = \frac{10 \text{ ohms}}{2} = 5 \text{ ohms}$$

You can now consider R_2 and R_3 to have the same effect as a

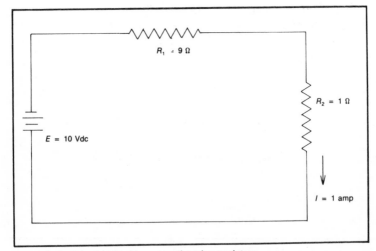

Fig. 3-8. Series circuit with unequal series resistors.

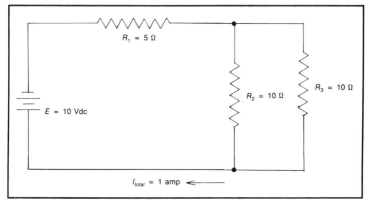

Fig. 3-9. Example of a series-parallel electric circuit.

single 5-ohm resistor in series with R_1, as shown in Fig. 3-10. Therefore, the total circuit resistance is:

$$R_{total} = R_1 + (5 \text{ ohms}) = (5 \text{ ohms}) + (5 \text{ ohms}) = 10 \text{ ohms}$$

This provides the information needed to calculate I_{total}:

$$I_{total} = \frac{E_{total}}{R_{total}} = \frac{10 \text{ volts}}{10 \text{ ohms}} = 1 \text{ amp}$$

The voltage dropped across R_1 can now be calculated as follows:

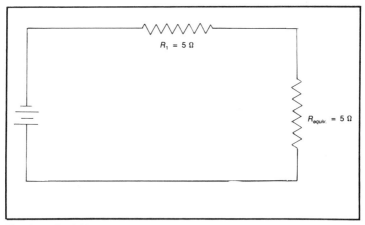

Fig. 3-10. Equivalent circuit of Fig. 3-9.

$$E = IR = (1 \text{ amp}) (5 \text{ ohms}) = 5 \text{ volts}$$

You know that the voltage across R_2 equals the voltage across R_3, because they are in parallel. You also know the voltage drop across the parallel network (R_2 and R_3) plus the voltage drop across R_1 must equal the source voltage of 10 volts. Therefore, the voltage dropped across R_2 and R_3 is also 5 volts. This is logical, because you calculated the equivalent parallel resistance of R_2 and R_3 to be 5 ohms, which is the same resistance value as R_1. Since R_1 and R_{equiv} are equal, you would expect the voltage to be divided equally.

Finally, you can calculate the currents flowing through R_2 and R_3. The current flowing through R_2 is:

$$I(R_2) = \frac{E(R_2)}{R_2} = \frac{5 \text{ volts}}{10 \text{ ohms}} = 0.5 \text{ amps}$$

You know that the resistance and voltage in R_3 are the same, so $I(R_3)$ will also equal 0.5 amp. If you add the currents flowing through R_2 and R_3, you should come up with the total circuit current as follows:

$$I(R_2) + I(R_3) = I_{total}$$
$$0.5 + 0.5 = 1 \text{ amp}$$

You had previously proven I_{total} to be 1 amp. This confirms your prior calculations.

POWER DISSIPATION

The last important factor to consider in an electrical circuit is *power dissipation*. Electrical power is dissipated (used up) in the form of heat.

It is often necessary to calculate the amount of power that must be dissipated in a circuit or component to keep from destroying something. Also, if you are supplying the power to a circuit, you need to know the amount of power to supply. You should memorize the following three equations to perform these calculations.

Power is equal to the voltage multiplied by the current:

$$P = E I \hspace{2cm} \textbf{Equation 3-4}$$

Power is equal to the current squared multiplied by the resistance:

$$P = I^2 R \qquad \textbf{Equation 3-5}$$

Power is equal to the voltage squared divided by the resistance:

$$P = \frac{E^2}{R} \qquad \textbf{Equation 3-6}$$

Examine how these equations could be used to calculate the power dissipated in Fig. 3-1. Assuming that you know the voltage and current in Fig. 3-1, you can use the following equation:

$$P = IE = (1)(10) = 10 \text{ watts}$$

If you know the current and resistance, you can use the following equation:

$$P = I^2 R = (1)(1)(10) = 10 \text{ watts}$$

If you know the voltage and resistance, you can use the following equation:

$$P = \frac{E^2}{R} = \frac{(10)(10)}{10} = \frac{100}{10} = 10 \text{ watts}$$

As you can see from the previous calculations, you only need to know two of the three circuit variables (current, voltage, and resistance) to calculate the power dissipation. If you wanted to know the power dissipated by a single component, you would use only those variables associated with that component. For example, you would calculate the power dissipated by R_1 in Fig. 3-3 as follows:

$$P = IE = (1)(10) = 10 \text{ watts}$$

The power dissipated by the entire circuit in Fig. 3-3 would be:

$$P_{total} = I_{total} \, E_{total} = (2)(10) = 20 \text{ watts}$$

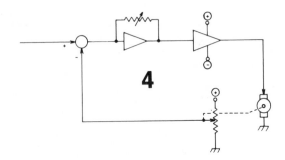

Algebraic Addition

A LTHOUGH THE TERM *ALGEBRAIC ADDITION* MAY SOUND
somewhat ominous, it is actually quite simple. In fact, you have
probably done it for years without realizing it.

In dealing with algebraic addition, you must consider the sign
of a number. In other words, you must determine whether the numbers in question are positive or negative.

When performing algebraic addition, positive and negative
numbers are simply combined. Consider the simple subtraction
problem 5 − 2 = 3. You were taught this form of basic math in
elementary school. But there are other approaches to the solution
of this problem.

Refer to Fig. 4-1, which shows a number line. The numbers
to the right of the zero are positive. The numbers to the left of the
zero are negative.

When dealing with a positive number, you must count that number of places to the right. When dealing with a negative number,
you must count that number of places to the left. The starting position will always be zero.

The previous subtraction problem indicates that you must count
five places to the right. (A number without a sign is always assumed
positive.) Thus, you will count to the + 5 position. Next, count two
places to the left, because the 2 has a negative sign. Thus, you will
count to the + 3 position, which is the correct answer. You actu-

Fig. 4-1. Shown is a number line. The numbers to the right of 0 are positive, and the numbers to the left of 0 are negative.

ally combined (algebraically added) a $+5$ and a -2, obtaining $+3$ as a result.

Consider what would result if you reversed the order of the numbers to be $-2 +5$. You would count two places to the left, arriving at the -2 position. Next, you would count five places to the right, arriving at the $+3$ position. Again, you obtained the same answer. When you algebraically add numbers, the order of the numbers makes no difference.

Similarly, you can algebraically add two negative numbers. For example, in the problem $-3 -2 = -5$, you are counting three places to the left and then counting two more places to the left. In total, you have counted five places to the left. For positive numbers, the same principle applies, except that you count to the right. For example, in the problem $+2 +1 = +3$, you count two places to the right and then one more place to the right, arriving at $+3$.

More than two numbers can be algebraically added at once. For example, in the problem $-2 +3 -4 +5 = +2$, you move a total of six places to the left ($-2 -4$) and eight places to the right ($+3 +5$).

EXAMPLE PROBLEMS

$$-20 +50 = 30 \qquad -78 -2 = -80 \qquad 56 -10 = 46$$
$$-56 +10 = -46 \qquad -10 -10 = -20 \qquad +3 +24 = 27$$
$$-480 -30 = -510 \qquad -36 +10 -6 -23 +89 = 34$$

You must understand the principle of algebraic addition before you proceed to the next chapter, because you must algebraically add exponents when performing simple Ohm's-law calculations on extremely large or small numbers.

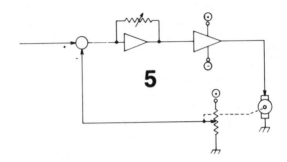

Scientific Notation

W HEN DEALING WITH THE BASIC UNITS OF ELECTRICAL MEA-
surement, you must often use extremely large and extremely
small numbers. As you learned in Chapter 1, commonly used
prefixes aid in simplifying these numbers. In review, these prefixes
are:

milli = 1/1,000   *pico* = 1/1,000,000,000,000
micro = 1/1,000,000   *kilo* = 1,000
nano = 1/1,000,000,000   *mega* = 1,000,000

Consider the following examples.
The value of 1 kilohm is 1000 ohms. Therefore:

4.7 kohm = 4700 ohms
1.5 kW = 1500 watts

The value of 1 megawatt is 1,000,000 watts. Therefore:
1.5 megaohm = 1500000 ohms
2.7 megawatt = 2700000 watts

The value of 1 milliamp is 1/1000 of an amp. (This can also
be expressed as .001 amp.) Therefore:

2.3 milliamps = .0023 amp
1.7 milliwatts = .0017 watt

The value of 1 microwatt is 1/1,000,000 of a watt. (This can also be expressed as .000001 watt.) Therefore:

$$6.8 \text{ microamps} = .000006 \text{ amp}$$
$$7 \text{ microvolts} = .000007 \text{ volt}$$

The value of 1 nanoamp is 1/1,000,000,000 of an amp. (This can also be expressed as .000000001 amp.) Therefore:

$$14 \text{ nanovolts} = .000000014 \text{ volt}$$
$$5 \text{ nanowatts} = .000000005 \text{ watt}$$

The value of 1 picowatt is 1/1,000,000,000,000 of a watt. (This can also be expressed as .000000000001 watt.) Therefore:

$$7 \text{ picoamps} = .000000000007 \text{ amp}$$
$$4.5 \text{ picovolts} = .0000000000045 \text{ volt}$$

The preceding examples show that prefixes are used for convenience. It is difficult to read or discuss .000000000003 amp, but it is easy to express 3 picoamps. Also, with a little practice, it becomes easy to conceive terms such as *milliamp* or *microwatt*. Another important point to understand is that each prefix is simply a means of expressing the true position of the decimal point.

The easiest way to mathematically use these prefixes (and other extremely large or small numbers) is by means of a simple method called *scientific notation*. Scientific notation is a means of expressing the position of a decimal point in terms of a base number and an exponent. Since the most commonly used numbering system is base 10 (decimal), the base number is always 10. The exponent describes the number of places to move the decimal point to the right or left to convert a decimal number to a whole number. If the exponent does not have a sign, it is assumed to be positive. Thus, the decimal point must be moved to the right. If the exponent has a minus ($-$) sign, it is known to be negative. Thus, the decimal must be moved to the left. For example:

$$1 \times 10^3 = 1000$$
$$1 \times 10^{-3} = .001$$
$$1 \times 10^2 = 100$$
$$1 \times 10^{-2} = .01$$
$$1 \times 10^1 = 10$$
$$1 \times 10^{-1} = .1$$

In the above examples, 10 is the base number, and the powers (i.e., third, second, first, etc.) are the exponents. The multiplier in all of the above examples is 1, but any number can be the multiplier. For example:

$$2.1 \times 10^2 = 210$$
$$30 \times 10^{-2} = .3$$

The prefixes you have learned can be used in scientific notation as follows

kilo = 10^3	*micro* = 10^{-6}
mega = 10^6	*nano* = 10^{-9}
milli = 10^{-3}	*pico* = 10^{-12}

The most confusing aspect of using scientific notation is determining the direction in which to move the decimal point. This depends upon whether you are converting a whole number to scientific notation or converting scientific notation to a whole number. Consider the following examples.

You may convert the number 5 to scientific notation as shown:

$$50 \times 10^{-1}$$
$$500 \times 10^{-2}$$
$$5000 \times 10^{-3}$$

All of the above expressions are equal to 5. Based on the known relationship between scientific notation and the commonly used prefixes, it can be said that 5 units is equal to 5000 milliunits. Consider what happens when it is necessary to convert in the opposite direction:

$$.5 \times 10^1$$
$$.05 \times 10^2$$
$$.005 \times 10^3$$

Again, all of the above expressions are equal to 5. Based on known relationships, it can be said that 5 units are equal to 0.005 kilounits. In other words, 5000 milliunits = 5 units = .005 kilounits.

Consider the following examples:

$$620 \text{ millivolts} = .62 \text{ volt} = .00062 \text{ kilovolt}$$
$$10{,}000 \text{ milliamps} = 10 \text{ amps} = .01 \text{ kiloamp}$$
$$1{,}000{,}000 \text{ milliohms} = 1{,}000 \text{ ohms} = 1 \text{ kilohm}$$

As can be seen from these examples, you may convert any number to virtually any prefix as long as you maintain the correct relative decimal point position. You must know how to perform these decimal-point conversions when using scientific notation in mathematical calculations.

There are three basic rules for using scientific notation in mathematics. If you are not familiar with using scientific notation, I suggest you memorize the following three rules and thoroughly understand the practice examples at the end of this chapter.

1. **Addition or Subtraction.** All expressions must have the same base number and exponent, or you must change them accordingly. The multipliers are then added or subtracted, and the common base number and exponent are placed at the end of the answer.
2. **Multiplication.** The base numbers must be the same. The exponents are algebraically added and the multipliers are multiplied.
3. **Division.** The sign of the exponent of the bottom expression (the divisor) is changed and algebraically added to the exponent of the top expression (the dividend). The base numbers must be the same. The multipliers are divided.

PRACTICE EXAMPLES

Example #1 (subtraction)

$$1.5 \text{ milliamps} - 0.4 \text{ milliamp} = ?$$

Convert the problem to scientific notation.

$$(1.5 \times 10^{-3}) - (0.4 \times 10^{-3})$$

The base numbers and exponents are the same, so simply subtract 0.4 from 1.5:

$$(1.5) - (0.4) = 1.1$$

Next, place the common base number and exponent at the end of the answer.

$$1.1 \times 10^{-3}$$

Finally, convert the scientific notation to the correct prefix.

1.1 milliamps

Example #2 (addition)

22.5 millivolts + 10.2 millivolts = ?

Convert the problem to scientific notation.

$$(22.5 \times 10^{-3}) + (10.2 \times 10^{-3})$$

To simplify this problem, you may eliminate the decimal point from the multiplier and reflect it in the exponent.

$$(225 \times 10^{-4}) + (102 \times 10^{-4})$$

The base numbers and exponents are the same, so simply add 225 and 102.

$$225 + 102 = 327$$

Next, place the common base number and exponent at the end of the answer.

$$327 \times 10^{-4}$$

Finally, you must convert the scientific notation back to the correct prefix. To convert back to *milli*, the decimal point must be moved 1 place to the left (*milli* = 10^{-3}). Therefore:

$$32.7 \times 10^{-3} = 32.7 \text{ millivolts}$$

At this point, you may be wondering why this last problem was not solved by the addition of 22.5 millivolts and 10.2 millivolts to obtain the correct answer of 32.7 millivolts, which would have been much simpler. It *is* easier to simply add the millivolt values to calculate the correct total, but consider the next example problem.

Example #3 (subtraction)

$$1.026 \text{ volts} - 33 \text{ millivolts} = ?$$

Convert the problem to scientific notation.

$$(1026 \times 10^{-3}) - (33 \times 10^{-3})$$

In order to obtain the exponents to the same value, 1.026 was converted to 1026×10^{-3}. Now you can simply subtract the multipliers.

$$1026 - 33 = 993$$

Next, place the common base number and exponent at the end of the answer.

$$993 \times 10^{-3} = 993 \text{ millivolts}$$

Finally, convert the answer to the desired prefix (or no prefix if applicable):

$$993 \text{ millivolts} = 0.993 \text{ volt}$$

Example #4 (multiplication)

$$P = IE$$

Assume voltage to be 5.2 volts. Assume current to be 12 milliamps. Therefore:

$$P = (12 \text{ milliamps}) (5.2 \text{ volts})$$

Convert the problem to scientific notation.

$$P = (12 \times 10^{-3}) (52 \times 10^{-1})$$

41

Next, multiply the multipliers.

$$(12) (52) = 624$$

Then, algebraically add the exponents.

$$-3 -1 = -4$$

This results in an answer of 624×10^{-4}. Finally, convert the answer to the most convenient prefix:

$$624 \times 10^{-4} = 62.4 \times 10^{-3} = 62.4 \text{ milliwatts}$$

Therefore, the final answer is 62.4 milliwatts.

Remember, whenever you multiply volts times amps, the answer must be in units of watts.

Example #5 (division)

$$R = \frac{E}{I}$$

Assume the voltage to be 50 volts. Assume the current to be 10 milliamps.

$$R = \frac{E}{I} = \frac{50}{10 \times 10^{-3}}$$

First, change (invert) the sign of the exponent of the bottom expression (divisor). Thus:

$$\frac{50}{10 \times 10^{3}}$$

Next, algebraically add the exponent of the top expression to the exponent of the bottom expression. (Because the dividend is not expressed in scientific notation, its exponent is 0.) Therefore:

$$3 + 0 = 3$$

At this point, you know that the base number and the exponent of your answer will be 10^3. Now, simply divide the multipliers.

$$\frac{50}{10} = 5$$

Now, place the base number and the exponent at the end of your answer.

$$5 \times 10^3$$

The expression 10^3 directly converts to the prefix *kilo*, so the final answer is 5 kilohms.

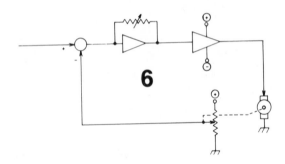

Basic Test Equipment

NOTE: The author strongly advises the reader not to attempt to use any test equipment until reading the equipment owner's manual and the section within this chapter entitled "Safety Considerations."

ONE OF THE MOST FREQUENTLY ASKED QUESTIONS BY THE maintenance technician is "What tools will I need?" This chapter deals specifically with this question.

The most important piece of test equipment you will need is the *digital voltmeter,* abbreviated DVM. However, a *volt-ohm-milliamp meter* (VOM) may be used in place of a DVM. These instruments are often commonly referred to as multimeters, or simply meters. Your DVM will eventually become an extension of your eyes and hands, so it is very important to know how to use it properly.

All DVMs and VOMs have the capability to measure the three primary circuit variables—voltage, resistance, and current. When you measure voltage or current, you must set the DVM or VOM for either ac or dc by selecting either ac or dc on a selector switch (or pushbutton). Resistance measurements are never taken with applied circuit power, so the terms ac and dc are irrelevant.

In this chapter, measurement of voltage, current, and resistance will be examined individually. If voltage measurements are being

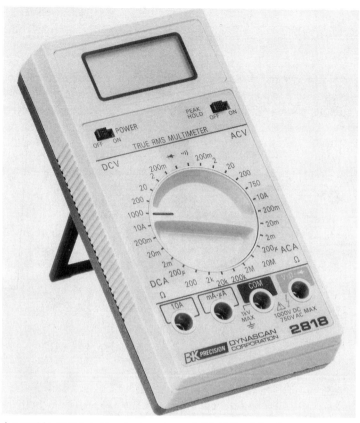

A portable digital multimeter (courtesy of B & K-Precision/Dynascan Corp).

discussed, the DVM or VOM will be referred to as a *voltmeter*; for resistance, it will be called an *ohmmeter*; and for current, it will be called an *ammeter*.

MEASURING VOLTAGE

Your voltmeter will contain various voltage ranges from which to select. It is important to develop the habit of checking the range switch before each voltage measurement to ensure that the correct range has been selected.

When you first examine your voltmeter, you will notice that the two test leads are red and black. The black lead is plugged into the connection on the voltmeter marked "common." As you learned in Chapter 1, all voltages are defined in respect to some common point. The black "common" lead is the lead connected to this reference point. The red lead is plugged into the connection on the volt-meter marked "volts." The voltmeter will read (or display) the voltage applied to the red lead in reference to the black "common" lead.

Voltage is measured by the placement of the voltmeter leads across the component to be measured. This is logical because voltage does not travel through anything; it is applied across components. To see it, you must place the test leads across the points in question.

A digital multimeter (courtesy of BBC-Metrawatt/Goerz, 2150 W. 6th Avenue, Broomfield, CO 80020).

NOTE: By placing your voltmeter across a component, it is important to realize that you are actually placing the voltmeter in parallel with that component. As you learned in Chapter 2, components in parallel will always have the same voltage across them. You prove this principle every time you use a voltmeter.

As with all types of meters, your voltmeter will also contain a switch for selecting either ac or dc voltage. When this switch is in the dc position, the voltmeter will only measure dc voltages and will not see ac voltages. The opposite is true if the switch is in the ac position.

Before taking a voltage measurement, you must choose the correct range setting for that measurement. There are two ways to determine the correct range:

1. If you are certain that you know the maximum value of

the voltage that you intend to measure, you may set the range switch accordingly.

2. If you are not sure of the maximum voltage you may see at the desired measurement point, you should set the voltmeter to its highest range for the first measurement. If your measured voltage is too low to read well at this range position, you can change the setting to a lower range position.

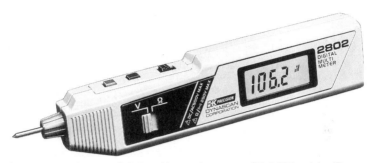

A compact probe-type digital multimeter (courtesy of B & K-Precision/Dynascan Corp.).

The previous procedure will keep you from "over-ranging" the voltmeter (applying a higher voltage to the test probes than the voltmeter is capable of reading at a specific range setting). Besides being potentially dangerous, this condition can result in severe damage to the voltmeter.

Most voltmeters can measure a maximum of approximately 1000 volts without special high-voltage adapters. Always read the owner's manual for your voltmeter to familiarize yourself with the voltmeter's limitations before attempting to use it.

Examine how you would check the voltages shown in Fig. 6-1 with a typical voltmeter. Obviously, all of the voltages within this circuit are dc, so you should set the voltmeter for dc volts. Your next consideration should be the desired range setting. In this case, you know the applied voltage, so you can pick the lowest range above that voltage. For example, if your voltmeter's ranges were 2, 20, 200, and 2000, you would pick the 20-volt range because it is the lowest range above the source voltage of 10 volts. (If you do not know the applied voltage, you should start at the highest voltage range and work your way down to keep from damaging the voltmeter.)

As shown in Fig. 6-1, the voltmeter is placed across R_2. The

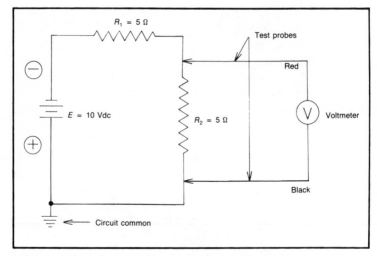

Fig. 6-1. Voltmeter being used to measure the voltage developed across a resistance.

voltage measured by the voltmeter is the voltage applied to R_2 in reference to the *circuit common*. Circuit common is the positive side of the battery. The symbol for circuit common is shown in Fig. 6-1 and Fig. 6-2. (Figure 6-2 shows two commonly used symbols for circuit common along with the symbol for earth ground.) The black test probe is placed on circuit common, and the red test probe is placed on the other side of R_2.

NOTE: In Fig. 6-1, the positive side of the battery, the lower side of R_2, and the point where the symbol for circuit common is shown

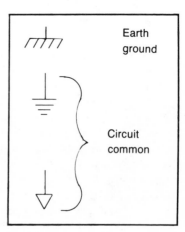

Fig. 6-2. Earth-ground and circuit-common symbols.

are all assumed to be the same electrical point. Under normal circumstances, this assumption is correct because the resistance of the circuit wires are negligible.

The voltmeter in Fig. 6-2 should measure – 5 volts dc. (You can prove this by using Ohm's law.) The negative voltage is due to the nature of the circuit common, or reference point, as discussed in Chapter 1. Since the positive side of the battery is the reference point, all other points in the circuit must be more negative than the reference point. If you reversed the test probes of the voltmeter, you would consider the top of R_2 to be the reference point. Your voltmeter would then measure + 5 volts dc.

NOTE: Some older styles of voltmeters are not capable of measuring negative voltages without reconnecting the test leads. If you have this type of voltmeter, simply reverse the test leads if the voltmeter deflects in the negative direction.

Another important consideration is the *input resistance* of the voltmeter. The owner's manual for your voltmeter may refer to the input resistance as *input impedance. Impedance*, a term used to define ac resistance, will be discussed in a later chapter. For now, you can simply consider impedance and resistance to be the same.

A good voltmeter should have an extremely high input resistance to reduce an undesirable effect called *loading*. Figure 6-3 demonstrates the result of using a faulty voltmeter to measure the same circuit you examined in Fig. 6-1. The voltmeter being used only has 5 ohms of input resistance. When you measure the voltage across R_2, you actually place the voltmeter in parallel with R_2. This causes the voltmeter input resistance to be in parallel with the circuit resistance. In this case, the 5 ohms of voltmeter input resistance will be placed in parallel with the 5 ohms of resistance of R_2. Thus, the equivalent resistance of the parallel network will only be 2.5 ohms. In other words, you will actually change the voltage across R_2 by trying to measure it.

Ohm's law can be used to examine this effect. First, consider how the circuit operates without using the voltmeter to measure the voltage across R_2. Since Fig. 6-3 (without the voltmeter) is a series circuit, R_1 and R_2 must be added to find the total circuit resistance. The circuit current (I_{total}) is the same through all components of the circuit. Therefore:

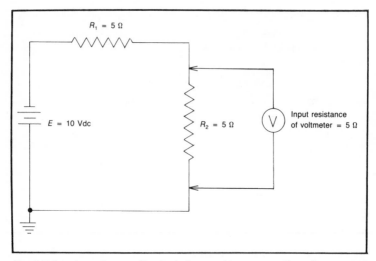

Fig. 6-3. Low-impedance voltmeter being used to measure the voltage across a resistance.

$$I_{total} = \frac{E_{total}}{R_{total}} = \frac{10 \text{ volts}}{R_1 + R_2} = \frac{10 \text{ volts}}{10 \text{ ohms}} = 1 \text{ amp}$$

Now, calculate the voltage across R_2.

$$E(R_2) = I_{total} R_2 = (1 \text{ amp}) (5 \text{ ohms}) = 5 \text{ volts}$$

Now, examine what occurs when faulty voltmeter (5-ohm input resistance) is placed across R_2. See Fig. 6-4. First, calculate the equivalent resistive effect of R_2 with the voltmeter in parallel. Because the value of R_2 (5 ohms) and the voltmeter input resistance (5 ohms) are equal, you can simply divide either value by two as follows:

$$R_{equiv} = \frac{5 \text{ ohms}}{2} = 2.5 \text{ ohms}$$

You can think of this 2.5 ohms as a single resistor in place of the parallel network of R_2 and the voltmeter. Therefore, you now have a simple series circuit. The total circuit resistance will be:

$$R_{total} = R_1 + R_{equiv} = 5 \text{ ohms} + 2.5 \text{ ohms} = 7.5 \text{ ohms}$$

50

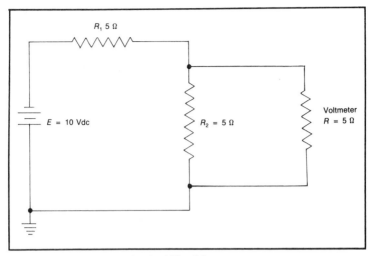

Fig. 6-4. Corresponding circuit of Fig. 6-3.

You can now calculate the total circuit current as follows:

$$I_{total} = \frac{E_{total}}{R_{total}} = \frac{10 \text{ volts}}{7.5 \text{ ohms}} = 1.33 \text{ amps}$$

The total circuit current increased when the faulty voltmeter was placed in the circuit. This is logical because the total circuit resistance decreased. Whenever resistance decreases in an operating circuit, the circuit current must increase.

So what effect did this have on the actual voltage measurement? R_{equiv} and I_{total} have already been calculated, so you can now calculate the voltage across E_{equiv} as follows:

$$E_{equiv} = I_{total} R_{equiv} = (1.33)(2.5) = 3.325 \text{ volts}$$

You can see the difference. As soon as you connected the voltmeter across R_2, you dropped the correct voltage value from 5 volts to 3.325 volts. This is called *loading*. It should be obvious from this example that accurate voltage measurements can only be obtained from a voltmeter with an extremely high input resistance (impedance).

Probably no manufacturer produces voltmeters with only 5 ohms of input impedance. But, on the other hand, you will probably need to measure voltage across much higher resistances than

5 ohms. If, for example, you needed to measure the voltage across a 1-megohm (1,000,000 ohms) resistor, the voltmeter must have an input impedance much higher than 1 megohm if an accurate reading is to be obtained.

MEASURING CURRENT

Your DVM (or VOM) will also be capable of measuring current. Like voltage, you must choose the correct range on your DVM, but you must also specify either ac or dc current. Typical current ranges available on most DVM's are from 2 mA (mA is the abbreviation for *milliamp*) to 2 amps. In this section, current measurements with a typical DVM (or VOM) will be examined.

An *ammeter* measures current. Since current flows through circuit components, you must actually insert the ammeter, as a series component, into a circuit to measure this flow. If the internal resistance of the ammeter is significant, it will drop voltage ($E = IR$). This added resistance and voltage drop will change the normal operating characteristics of the circuit, and the corresponding current reading will be inaccurate. If, on the other hand, the ammeter's input resistance is insignificant, it will simply replace the interconnecting wire in the circuit. For this reason, the input (or internal) resistance of a good ammeter should approach zero. Note this is exactly the opposite of a good voltmeter.

The first step in measuring current with an ammeter is to turn off all applied power to the circuit in question. You must physically disconnect the portion of the circuit for which you will measure the current flow.

When using a DVM or a VOM as an ammeter, you must unplug the red test lead from the volts-ohms connection and plug it into the connection labeled amps or milliamps. When this has been accomplished, the black test lead should be connected to the negative side of the open circuit, and the red test lead should be connected to the positive side. See Fig. 6-5. Set the range switch (or pushbutton) to the highest current range possible.

NOTE: DVM's or VOM's are the most susceptible to damage while trying to measure current. Always double-check the circuit to make certain that the circuit current cannot exceed the ammeter range before circuit power is applied.

Apply power to the circuit. If the ammeter measures negative

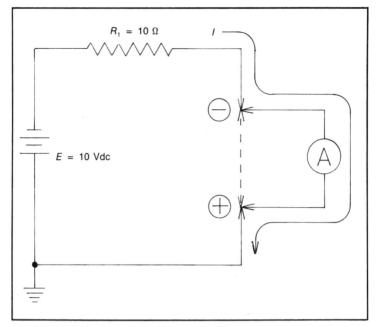

Fig. 6-5. Method of measuring dc current with an ammeter.

current, you have connected the ammeter test leads in the wrong polarity. If you are using a DVM as an ammeter, this will probably not present a problem, because most DVM's will measure equally well in either direction.

On the other hand, if you are using a VOM as an ammeter, you may be trying to drive the indicating needle the wrong way. Turn off the circuit power, and reverse the test leads. If your current range is too high to accurately measure a small current, switch to a lower range for a more accurate reading. The most accurate reading can be obtained at the lowest range possible without over-ranging the ammeter.

Figure 6-5 shows an ammeter being used to measure the current in a simple circuit. The dotted line represents the wiring of the original circuit before it was disconnected to insert the ammeter. Note that the entire circuit current must flow through the ammeter. The red lead of the ammeter should be connected to the point marked (+) and the black lead to the point marked (–).

Assuming that the ammeter has negligible internal resistance, the correct circuit current can be calculated through the use of Ohm's law.

$$I = \frac{E}{R} = \frac{10 \text{ volts}}{10 \text{ ohms}} = 1 \text{ amp}$$

The ammeter reading should show 1 amp of current flow. Now consider the effect of a significant amount—say 10 ohms—of internal resistance within the ammeter. It should be obvious that the resistance of R_1 and the internal ammeter resistance are in series. Therefore, the total circuit resistance is:

$$R_{total} = R_1 + R_{internal} = 10 + 10 = 20 \text{ ohms}$$

The circuit current can now be calculated as follows:

$$I = \frac{E}{R} = \frac{10 \text{ volts}}{20 \text{ ohms}} = 0.5 \text{ amp}$$

By adding 10 ohms of internal ammeter resistance, the current flow of the circuit was decreased from its normal 1 amp to 1/2 amp. The previous example demonstrates why an ammeter must have a very low internal resistance for accurate current measurements.

MEASURING RESISTANCE

When using a DVM or VOM to measure resistance, the first step is to make certain that all circuit power is off before any measurements are attempted. Resistance is *never* measured with the circuit power applied. The red test lead must be plugged into the ohmmeter connection labeled "ohms" or "resistance." The black test lead should remain in the common connection.

As with voltage and current measurements, the proper resistance range should be chosen for obtaining an optimum measurement. In the case of measuring resistance, you cannot damage anything by attempting the measurement while in the wrong range, but your reading may not be accurate. As in the case of measuring voltage and current, the correct range should be as low as possible without over-ranging the ohmmeter.

For example, assume that you are using an ohmmeter with the following ranges: 200 ohm, 2 kohm, 20 kohm, and 200 kohm. You want to measure the resistance of a resistor labeled 1 kohm (1,000 ohms). At the 200-kohm range, your reading will be 1. At the 20-kohm range, your reading may change to 1.1. At the 2-kohm range, your reading may change to 1.13. At the 200-ohm range, the ohmmeter will not register a measurement, because the 1-kohm resistor has a greater resistance value than the 200-ohm range can measure. Obviously, the most accurate measurement in this example was taken in the 2-kohm range. (A reading of 1.13 kohms is more precise than the 1-kohm reading obtained in the 200-kohm range.)

The labeling of the resistance ranges may vary somewhat between DVMs and VOMs or from one manufacturer to another. For example, a VOM may describe its different resistance ranges in terms of $R \times 1, R \times 10, R \times 100, R \times 1\,K$, etc. If you have any questions regarding the specific instrument you intend to use, read the owner's manual for that instrument or contact the manufacturer.

OTHER COMMONLY USED INDUSTRIAL TEST EQUIPMENT

A *current-transformer* ammeter (commonly known by the manufacturer's trade name Amprobe) is an ammeter for measuring high values of ac current. This type of ammeter will not measure dc currents.

The front of this instrument consists of a large pair of spring-loaded tongs to be opened and placed around the wire through which current flow is to be measured. The circuit does not have to be opened, and no physical contact needs to be made to the wire. This instrument actually measures the strength of the moving magnetic field created around any wire through which ac current flows. It converts this field strength reading to a proportional current reading.

Also available is a variety of clamp-on ammeters for measuring high values of dc current. These instruments are based on the Hall-effect principle, which will be discussed in a later chapter.

The *oscilloscope* can provide a visual representation of voltage variations (commonly called *waveforms*) within an operating circuit. In addition to displaying these waveforms, the oscilloscope can also be used to measure their voltage amplitude and frequency.

In many ways, an oscilloscope is similar to a small television set. The *cathode-ray tube* (picture tube) is used for displaying the voltage or current waveforms. The waveform amplitude is calculated by measuring the vertical height of the waveform and comparing it to the vertical sensitivity adjustment on the front panel of the oscilloscope.

The frequency of the waveform is calculated by measuring the horizontal length of one complete waveform (one complete cycle)

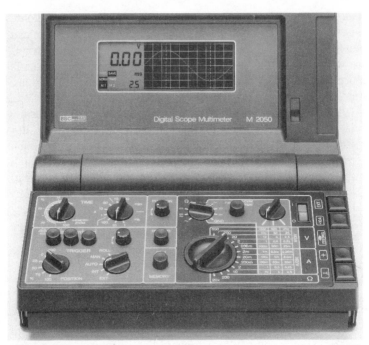

A portable digital oscilloscope and multimeter combination (courtesy of BBC-Metrawatt/Goerz, 2150 W. 6th Avenue, Broomfield, CO 80020).

and comparing it to the horizontal-sweep frequency adjustment on the front panel of the oscilloscope. If this seems confusing, don't worry. This brief description is only meant to acquaint you with the basic operating concept of an oscilloscope. To understand how to properly use an oscilloscope, read an oscilloscope user's man-

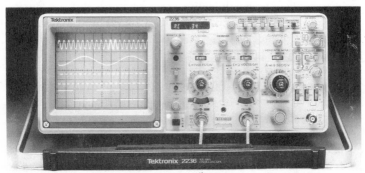

A modern dual-channel oscilloscope with multiple functions (courtesy of Tektronix, Inc.).

ual or a good book on the subject. Also, as you progress further in this book, the concept of waveform analysis will become much clearer.

A *megger* is used for checking the integrity of high-voltage insulation (such as that used in industrial electric motors). Many types of electrical shorts (undesired current paths) and insulation breakdowns can be detected with a megger. But do not use a megger in conjunction with solid-state electronic equipment. The high voltage created by the megger can destroy solid-state devices.

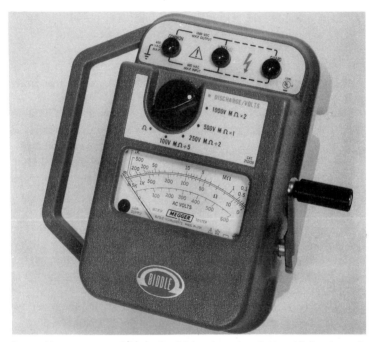

A portable megger used for testing high-voltage breakdown (dielectric quality) (courtesy of Biddle Instruments, Blue Bell, PA 19422).

SAFETY RULES TO FOLLOW
WHEN USING TEST EQUIPMENT

When using test equipment, you must adhere to the following precautions, *especially when working with or near potentials above 28 volts.*

1. Read and thoroughly understand the owner's manual supplied with the test equipment you intend to use before at-

tempting to use it. It is important to understand the equipment's proper uses and limitations.

2. Most electrical fatalities are the result of excessive electrical current passing through the body's vital organs (primarily the heart). For this reason, *always use only one hand* to perform tests or to take measurements on high-voltage equipment (over 28 volts) while power is applied. (This is commonly called the one-hand method.) Never touch anything conductive with the unused hand while tests or measurements are being performed. Also, do not stand on conductive material such as water or metal grating. If only one hand is near high voltage, a closed circuit through your vital organs cannot occur if the remainder of your body is isolated from conductive material.

3. Do not attempt to catch falling objects while near high-voltage equipment.

4. When using line-powered test equipment (test equipment which obtains its operating power from standard 120 Vac household power), always isolate it with an isolation transformer before attempting any measurements. Isolation transformers can be purchased from most electronics firms that sell test equipment.

5. Disconnect all circuit power before opening the circuit to insert an ammeter for measuring current. Apply the circuit power only after the ammeter is securely in place.

6. When you measure voltage or current, always start on the highest range of the meter and then proceed to the lower ranges to prevent damage to the meter.

7. Never attempt to measure resistance with any power applied to the circuit or component in question.

8. Follow all safety recommendations set forth in the user's manual supplied with the test equipment being used.

FINAL CONSIDERATIONS

Test equipment is expensive. With a little ingenuity and a good DVM, almost any type of troubleshooting is possible. This is not to imply that other test equipment (such as oscilloscopes) is not needed. However, you should not assume that the measure of a troubleshooter's ability is in the equipment possessed by the troubleshooter. (In addition to the cost, it is difficult to carry a complete electronic laboratory from place to place.)

A triple output bench power supply used for testing dc-powered equipment (courtesy of B & K-Precision/Dynascan Corp.).

In the following chapters, you will learn more about simple troubleshooting procedures that can be performed with a bare-bones budget. Review these procedures before selecting your test equipment.

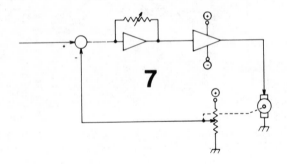

Resistors

BEFORE PROCEEDING ANY FURTHER, IT IS NECESSARY TO DE-fine a few common terms:

Component. A single device, or part, used in an electronic circuit.

Discrete component. A nonintegrated component. (Integrated circuits will be detailed in a later chapter.)

Discrete circuit. An electronic circuit composed entirely of individual discrete components.

The *resistor* is the most common discrete electronic component used in industrial electronics. As the name implies, the primary purpose of a resistor is to provide resistance (opposition) to current flow. The symbol for a resistor is shown in Fig. 7-1. Resistance will always be expressed in ohms.

Resistors have many common uses. For example, resistors limit the current flow of a circuit to some safe or desirable level. Like a valve in a water line, a resistor limits the maximum possible flow.

According to Ohm's law ($E = IR$), any resistive device with current flowing through it will develop a voltage across it. This is commonly referred to as the *voltage drop*. For example, if a resistor in a circuit develops 5 volts across it, it is said to *drop* 5 volts. Because this voltage drop can be accurately calculated (using Ohm's law), resistors are commonly used to set up desired operating

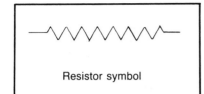

Fig. 7-1. Resistor symbol.

Resistor symbol

voltages. By using the proper resistor values, the source voltage can be tailored to the needs of other components within an electronic circuit. The tailored voltage is often called a *bias*.

When multiple resistors are connected in series to a source, as in Fig. 7-2, the source voltage will be divided proportionally to the individual resistor values. A circuit connected in this manner is commonly called a *voltage divider*.

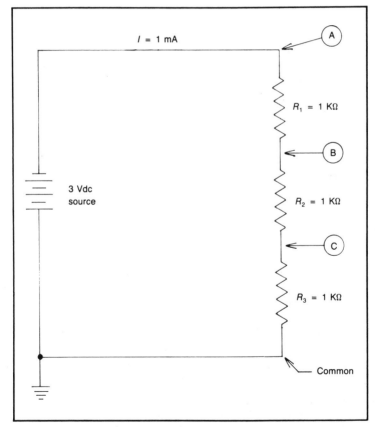

Fig. 7-2. Example of a voltage divider.

By using Ohm's law, the current in the circuit in Fig. 7-2 is calculated to be 1 milliamp (abbreviated mA). Since this is a series circuit, the amount of current flow will be the same through all three resistors. Therefore, the voltage drop across each resistor can be calculated (using Ohm's law) to be 1 volt. By using a voltmeter with the common lead connected to the common point (reference), you can measure three different voltages at points A, B, and C. The voltage at point A will be 3 volts because you are measuring directly across the 3-volt source. The voltage at point B will be 2 volts because of the 1-volt drop across R_1. Finally, the voltage at point C will be 1 volt because of the voltage drop of R_1 (1 volt) and R_2 (1 volt). Observe that all of these voltages are negative because circuit-common is the positive side of the battery.

Another way of looking at this is to consider the voltage drops in a series circuit to be additive. When you measure the voltage at point C, you are actually measuring across R_3 (1 volt). When you measure the voltage at point B, you are measuring the 1 volt drop of R_3 in addition to the 1 volt drop of R_2. This gives you a total of 2 volts. Finally, the voltage at point A will be the sum of the voltage drops of R_1, R_2, and R_3; or 3 volts. Therefore, this circuit is capable of supplying three different voltages (1, 2, or 3 volts) from only one source.

NOTE: This is a good time to refresh your memory concerning one of the basic rules discussed in Chapter 1: the individual voltage drops across the individual components in a series circuit must equal the source voltage.

In Fig. 7-2, resistors of equal value were used. Therefore, the voltage drops likewise had to be equal. Conversely, if unequal resistor values are used, unequal voltage drops will develop. (The voltage will divide proportionally to the resistance value.)

The circuit in Fig. 7-3 demonstrates this principle. In this circuit, four different resistor values are used, resulting in four different voltage taps (measurement or connection points). Remember, all of these voltages are in reference to a common point. For practice, try using Ohm's law to prove that these voltages are correct.

In summary, resistors are commonly used to provide the necessary operating voltages for other devices in electronic circuits. Resistors are also used to limit excessive or destructive current flow.

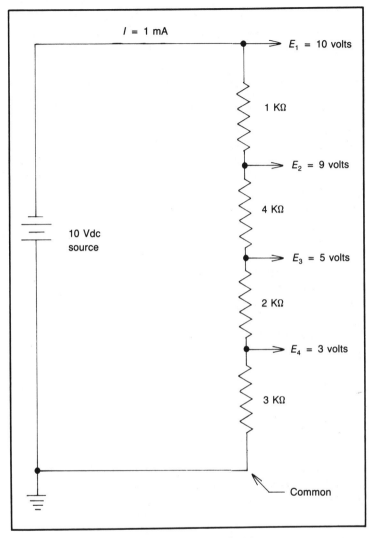

Fig. 7-3. Voltage divider incorporating unequal resistors.

RESISTOR TYPES

Most resistors are composed of carbon composition. They utilize the poor conductive characteristics of carbon to provide resistance. Other types of commonly manufactured resistors are carbon film, metal film, molded composition, thick film, and vitreous enamel. These different types possess various advantages (or disadvantages) related to such characteristics as temperature stability, tolerance, power dissipation, noise characteristics, and cost.

Temperature stability is a term describing how the operating characteristics of an electronic component will change with a corresponding change in temperature. In the case of resistors, this term is associated with a change in resistance relating to a change in the resistor's temperature.

Tolerance describes the allowable deviation in actual resistive value from the manufacturer's stated value.

Power dissipation describes the maximum allowable power the resistor can dissipate (use up in the form of heat) and still maintain reliable operation.

In some extremely sensitive electronic circuits, the internal electrical noise generation of a resistor becomes a critical factor. In these cases, the noise characteristics of resistors may be specified.

Special power resistors are used in applications requiring the resistor to dissipate large quantities of power (heat). These resistors are designed to operate at high temperatures and are usually constructed of high-resistance wire and ceramic.

Resistor arrays are also available. A resistor array consists of multiple resistors (usually of the same value) in one package. The package style can be a dual in-line package (DIP) style or any one of numerous other packaging styles.

RESISTOR COLOR CODES

Most resistors are surrounded by colored bands that define their resistance values. In order to repair or troubleshoot electronic equipment, you must memorize this color code system so that you will know which resistance values you are working with. Following are the color codes:

Black = 0	Green = 5
Brown = 1	Blue = 6
Red = 2	Violet = 7
Orange = 3	Gray = 8
Yellow = 4	White = 9

The color band closest to the end of the resistor represents the first digit of its resistance value. The second band from the end represents the second digit. The third band from the end represents the multiplier. (The multiplier simply states how many zeros are added to the first two digits.) The fourth band from the end is the *tolerance band*, which describes how close the actual value is to the stated value.

The absence of a tolerance band indicates the resistor has a 20 percent tolerance. In other words, the actual resistance value may be 20 percent higher or lower than its stated value. A silver band indicates a 10 percent tolerance and a gold band indicates a 5 percent tolerance.

On precision resistors or large power resistors, the actual value and the tolerance may be printed on the resistor body.

EXAMPLE: A resistor has four color bands. The first is red; the second, violet; the third, orange; and the fourth, silver. Thus, the first digit is 2; the second, 7; the third, or the multiplier, is 3; and the tolerance is 10 percent. The value of this resistor is 27,000 ohms with a possible 10 percent error. It can also be expressed as a 27 kohm resistor.

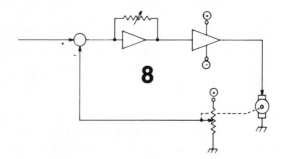

8

Potentiometers, Rheostats, and Resistor Banks

I N MANY APPLICATIONS, A VARIABLE RESISTIVE DEVICE IS needed to change certain operating parameters of an electronic circuit. For example, the volume control on a typical radio is a variable resistive device; it controls the volume level (or amplitude) being output from the radio's audio circuits. A radio's volume control is also a continuously variable device; as the volume control is rotated, the adjustments are continuous rather than in steps. In effect, an infinite number of adjustments can be made from the maximum to minimum levels.

POTENTIOMETERS

A *potentiometer*, as shown in Fig. 8-1, is a three-lead continuously variable resistive device. The two connection points labeled 1 and 3 are fixed resistances. The connection point labeled 2 is a variable tap, which is mechanically attached to a shaft capable of moving the tap from one end of the fixed resistance to the other. The shaft is usually rotated to provide the tap movement, but many special potentiometers are manufactured to move the tap in an up-and-down or in-and-out movement.

The actual resistive value from the tap to either side of the fixed resistance depends upon its physical position relative to either side of the fixed resistance. For example, in Fig. 8-1, if the tap is positioned at the top extreme (very close to point 1 of the fixed resis-

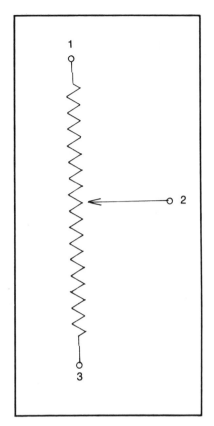

Fig. 8-1. Potentiometer symbol.

tance), the resistance value from point 1 to the tap will be very low. At the same time, the resistance value from point 2 to the tap will be very high. If the tap is positioned near the middle of the fixed resistance, the resistance value from the tap to point 1 or point 2 will be about equal (approximately 1/2 the total value of the fixed resistance). Likewise, if the tap is positioned near the bottom of the fixed resistance, the resistance. value from the tap to point 3 will be very low, and the resistance value from the tap to point 1 will be very high. As the tap is moved from one extreme to the other, the number of possible resistance values obtainable is infinite.

Note that as the tap of a potentiometer is varied, the two resistance values created between the tap and the fixed resistance points will vary in opposite directions. In other words, as one value increases, the other value decreases, and vice versa.

The primary specification of a potentiometer is the value of the fixed resistance. For example, if you specified a 1-kohm potentiometer, the fixed resistance of this potentiometer would be approximately (depending on its tolerance value) 1 kohm. By adjusting the tap position, you could vary the resistance of the tap to either side of the fixed resistance from 0 ohms to 1 kohm.

In the circuit shown in Fig. 8-2, one side of the fixed 1-kohm resistance is connected to a +10 Vdc source, and the other side is connected to the circuit common. A high-input impedance (resistance) voltmeter has been connected to the adjustable tap. (A high-input impedance voltmeter is desirable to minimize the loading effect, as discussed in Chapter 6.)

If the adjustable tap is in position 1, you should measure approximately 10 volts, because there is virtually no resistance between the source and the tap. As the tap is adjusted down to position 2 (approximately half of the fixed resistance), you will measure about 5 volts, because a voltage divider has been formed (consist-

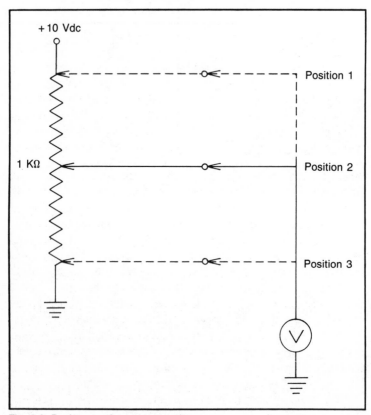

Fig. 8-2. Basic potentiometer operation.

ing of the resistance from the tap to the source and the resistance from the tap to the common point). If these two resistance values are equal, the tap voltage should be one-half of the source voltage. If the tap is adjusted down to the bottom extreme, you should measure approximately zero volts, because the tap is essentially connected to the circuit common.

In the preceding example, a continuously variable voltage divider has been demonstrated. In other words, any desired voltage from 10 volts down to zero can be tapped off. Likewise, if some form of electronic signal was connected in place of the 10-volt source voltage, you could tap off any percentage of the electronic signal desired.

All potentiometers have power ratings that must not be exceeded, regardless of the setting of the potentiometer. The power dissipated in the circuit shown in Fig. 8-2 is:

$$P = \frac{E^2}{R} = \frac{(10)\,(10)}{1000\ \text{ohms}} = \frac{100}{1000} = 0.1\ \text{watt}$$

The parallel resistance of the voltmeter does not have to be considered, because the voltmeter has a high-input impedance. The loading effect of the voltmeter is considered negligible. But consider the circuit shown in Fig. 8-3. In this illustration, a significant load has been connected to the potentiometer tap. The potentiometer is adjusted so 5 volts are dropped across the 5-ohm load resistor. Based on this information, the following statements can be made:

- A series-parallel circuit exists. The parallel portion consists of the load resistor in parallel with the resistance from the potentiometer tap to the common point. The series portion consists of the resistance from the potentiometer tap to the 10-volt source. Figure 8-4 illustrates this series-parallel circuit.
- If one-half of the source voltage is dropped from the potentiometer tap to the common point (across the load resistor), then one-half of the source voltage must be dropped from the potentiometer tap to the source.
- If the two voltage drops (from tap to source and across the load resistor) are equal, the two resistance values must also be equal, because the voltage drops will be proportional to the resistance. This means the resistance from the tap to the

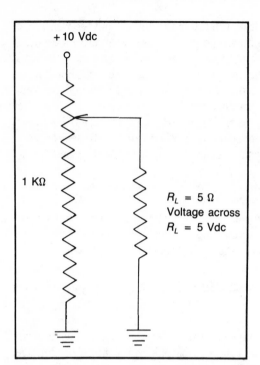

+ 10 Vdc

1 KΩ

$R_L = 5\ \Omega$
Voltage across
$R_L = 5$ Vdc

Fig. 8-3. Potentiometer with a load connected to the wiper tap.

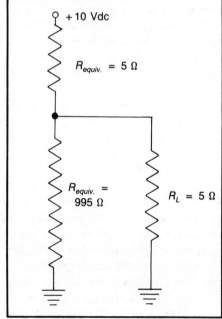

○ + 10 Vdc

$R_{equiv.} = 5\ \Omega$

$R_{equiv.} = 995\ \Omega$

$R_L = 5\ \Omega$

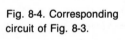

Fig. 8-4. Corresponding circuit of Fig. 8-3.

source must be approximately equal to the 5-ohm load resistor. (The 995-ohm resistance of the potentiometer in parallel with the 5-ohm load resistor can be considered to have a negligible effect. To prove this, you can calculate the equivalent resistance of 995 ohms and 5 ohms in parallel.)

Based on the previous statements, the resistance of the potentiometer from the tap to the source can be assumed to be 5 ohms. Using this 5-ohm resistance and the known voltage drop of 5 volts across the resistance, the power dissipation can be calculated as follows:

$$P = \frac{E^2}{R} = \frac{(5)\,(5)}{5\ \text{ohms}} = \frac{25}{5} = 5\ \text{watts}$$

With a very high resistance (a voltmeter) connected from the tap to the common point, the circuit in Fig. 8-2 only dissipates 0.1 watt. With a low resistance connected from the potentiometer tap to the common point, the power dissipation can increase dramatically, depending on the position (adjustment) of the tap. If, for example, the potentiometer was adjusted so the resistance from the tap to the source was a value other than 5 ohms, the power dissipation of the potentiometer would change drastically. (To prove this, calculate the power dissipation of the potentiometer adjusted to its mid-position.)

In summary, the power dissipation of a potentiometer will depend upon:

1. The load as seen by the potentiometer tap.
2. The voltage (or signal) level applied to the potentiometer.
3. The position of the potentiometer tap.
4. The fixed-resistance value of the potentiometer.

Another type of potentiometer is commonly called a *trim potentiometer*. A trim potentiometer is a potentiometer usually meant to be installed in a small space or on a printed circuit board. The power ratings are usually low (1 watt or less), and the adjustment is typically made with a small screwdriver. Trim potentiometers are used in applications requiring a one-time adjustment to compensate for circuit variables that cannot be precisely calculated. For example, many modern television sets require the use of trim potentiometers for fine-tuning local stations. Once this fine-tuning has been accomplished, there is no need for further adjustments unless the television set is moved to a different area.

Potentiometers may be specified as either *linear* (proportional change in resistance to a proportional change in tap position) or *logarithmic* (an exponential relationship between tap position and resistance value). In the case of a linear potentiometer, a 25 percent movement of the tap will result in a 25 percent change in resistance. The same change in tap position on a logarithmic potentiometer may result in a 90 percent change in resistance.

Logarithmic potentiometers are sometimes called *audio taper* potentiometers. This is because logarithmic potentiometers are often used in audio applications. The sensitivity to volume level of the human ear is nonlinear. Logarithmic potentiometers closely approximate this nonlinear sensitivity and are used for most audio volume-level controls. Logarithmic potentiometers are also useful in many other nonlinear applications.

RHEOSTATS

If one side of a potentiometer's fixed resistance is tied to the variable tap, the potentiometer becomes a *rheostat*, or continuously variable resistor. Figure 8-5 shows the electrical symbols for rheostats. Rheostats can be constructed from potentiometers or obtained as dedicated rheostats. Some rheostats are designed for high-power applications and are specially fabricated to dissipate large amounts of heat.

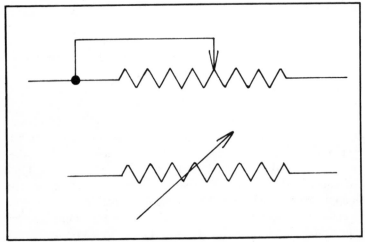

Fig. 8-5. Rheostat symbols.

ADJUSTABLE RESISTORS

Adjustable resistors should not be confused with potentiometers or rheostats. Potentiometers and rheostats are used in applications requiring frequent or precise adjustment. Adjustable resistors are meant to be adjusted once (usually to obtain some hard-to-find resistance value). The body of an adjustable resistor is similar to that of a typical power resistor except that it also contains a metal ring that can be moved back and forth across the resistor body. The position of this metal ring determines the actual resistance value. Once the desired resistance value is obtained, the metal ring is permanently clamped in place.

RESISTOR BANKS

Sometimes a resistor bank (consisting of multiple power resistors) is fabricated for custom applications and possibly to dissipate a large amount of power (heat). Frequently, resistor banks also incorporate multiple switching arrangements so the equivalent resistance may be easily (or automatically) changed in steps, or increments. Variable resistor banks are sometimes called *variable loads, dummy loads,* or *power loads.*

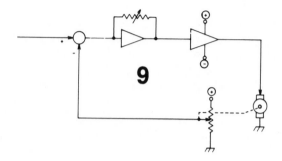

9

Alternating Voltage and Current

AS DESCRIBED BRIEFLY IN CHAPTER 1, THE TERM *ALTERNA-ting current* (abbreviated ac) means the current flow periodically changes direction. This condition is caused by periodic changes in the polarity of the applied voltage. Alternating voltage and current is very important because of a phenomenon known as *transformer action*. Transformer action enables the efficient transmission of large quantities of power over long distances. This is why all common household current is ac. (Transformer action will be discussed in a later chapter.) This chapter deals with some of the unique characteristics of alternating current.

AC WAVESHAPES

Usually, alternating voltages or currents have waveshapes. Common household ac has a waveshape called a *sine wave* (short for sinusoidal wave). This type of waveshape is shown in Fig. 9-1. A true sine wave can be thought of as a circle in which the top and bottom halves are separated. Referring to Fig. 9-1, if the top half of the waveshape were moved directly over the bottom half, it would form a circle. The sine wave is the most common type of ac waveshape used for any purpose.

If you have not been previously exposed to the concept of a waveform, or waveshape, this principle may seem complicated, but it is actually simple. A waveform is a graphic representation of the

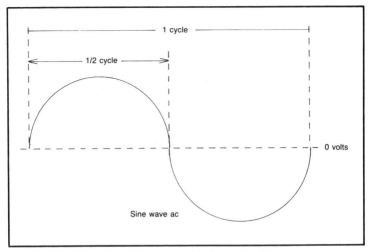

Fig. 9-1. One cycle of a sinusoidal (or sine) ac waveshape.

changing level (or amplitude) of voltage or current in relation to time.

For example, Fig. 9-2 is a graph representing the level change in a water tank over a period of about 20 hours. Every hour, the actual level at that point in time is measured, and a corresponding dot is plotted on the graph. The water level is represented on the vertical plane, while time is represented on the horizontal plane. During periods of heavy water usage, the level drops to a low peak. During periods of light water usage, the level reaches a high peak. If the water usage occurs in a repeatable, or predictable, manner, the level in the tank will also vary in a repeatable manner. This type of repeatable variation in water level is called *cyclic variation*; that is, its variations are repeatable. One cycle represents the total variation from some earlier point in time to the point at which the variation begins to duplicate itself.

The graphical representation of Fig. 9-2 is analogous to the electrical sine wave shown in Fig. 9-1. Time is always represented in the horizontal plane, and amplitude in the vertical plane. In this manner, you can accurately define variations in the amount of voltage or current and the amount of time involved in these variations. Any cyclic condition can be defined in this manner.

The level of water in a tank does not always vary in a cyclic manner. This is also true for voltage or current. If any variable (water level, voltage, current, or anything capable of changing) does

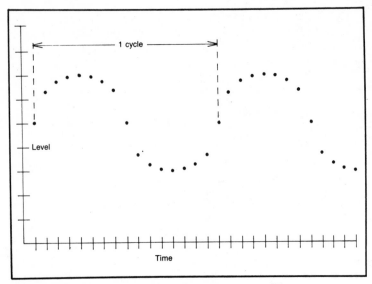

Fig. 9-2. Graphical analogy of a cyclic process or condition.

not change in a repeatable manner, it is said to be *random*. Random voltages and currents cannot be measured in terms of waveshapes or time periods.

AC FREQUENCY

Figure 9-1 illustrates one cycle of a typical ac sine wave. One cycle is defined as one complete periodic variation. In the case of common household ac, these periodic variations occur at the rate of 60 times per second. This is commonly called 60 *hertz* (cycles per second). The rate of periodic variations is called *frequency*. Common household ac power has a frequency of 60 hertz.

There is an inversely (opposite) proportional relationship between frequency and the time period of one cycle. Equations 9-1 and 9-2 show how time is calculated from frequency and frequency from time.

$$\text{Time period} = \frac{1}{\text{frequency}} \quad \textbf{(Equation 9-1)}$$

$$\text{Frequency} = \frac{1}{\text{time period}} \quad \textbf{(Equation 9-2)}$$

For example, if you wanted to calculate the time period for one cycle of 60 hertz ac, you would simply calculate the reciprocal of the frequency. As you have learned, the reciprocal of any number is found by dividing the number into one. Thus:

$$\text{Time period} = \frac{1}{\text{frequency}} = \frac{1}{60} = .0166 \text{ second}$$

To simplify the above answer, express it as 16.6 milliseconds.

Likewise, if you knew the time period of one cycle and wanted to calculate the frequency, you would simply find the reciprocal of the time period. For example, if the time period is 16.6 milliseconds:

$$\text{Frequency} = \frac{1}{\text{time period}} = \frac{1}{.0166} = 60 \text{ hertz}$$

AC AMPLITUDE

One-half of a complete cycle is appropriately called a *half-cycle*. As shown in Fig. 9-1, for each half-cycle, the voltage or current actually passes through zero. In fact, it *must* pass through zero to change polarity or direction. If Fig. 9-1 is a voltage waveform, the voltage is shown to be positive during the first half-cycle and negative during the second half-cycle.

The horizontal dotted line representing zero voltage is called the *zero reference line.* As shown in Fig. 9-3, the maximum amplitude, as measured from the zero reference line, is called the *peak voltage.* The total deviation from the negative peak (the peak below the zero reference line) to the positive peak (the peak above the zero reference line) is called the *peak-to-peak voltage.*

Defining the level of pure dc voltages or currents is easy, because the dc level is constant and continuous. Defining ac voltage or current levels is a little more complicated. As shown in Fig. 9-3, the peak voltage of this waveform is 10 volts. This represents an instantaneous voltage level. Most of the time, the voltage is less than 10 volts (sometimes even zero) and can be of the positive or negative polarity. For this reason, there are numerous ways to accurately define alternating voltage and current levels.

Often, you can define the ac level (amplitude) by algebraically adding the negative half-cycle to the positive half-cycle. In Fig. 9-3,

77

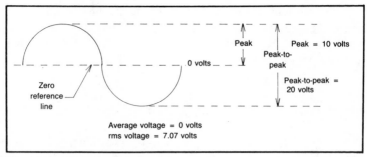

Fig. 9-3. Relationship of peak, peak-to-peak, average, and rms voltage relative to a sinusoidal waveshape.

the negative half-cycle cancels out the positive half-cycle, resulting in an average voltage of zero (average current would also be equal to zero). This is true for all sine-wave ac voltages. If this is difficult for you to understand, consider this analogy. A dc voltage and current is like driving a car in one direction. The passengers in this car will go somewhere. In contrast, an ac voltage and current is like driving a car back and forth in a driveway. The passengers in this car can ride all day long and still end up where they started. This is why a dc voltmeter will measure 0 volts if a pure 60 hertz ac voltage is applied to the test probes. The dc voltmeter cannot respond to the rapid ac polarity reversals. Therefore, it indicates the average level, or zero.

If the ac voltage shown in Fig. 9-3 was applied to a load (as shown in Fig. 9-4), current would flow in a back-and-forth direction, proportionally following the voltage variations. According to Ohm's law, some quantity of power would be dissipated, because current flows through the load. The power dissipated while the current flows in one direction does not negate the power dissipated when the current reverses (there is no such thing as negative power). Therefore, power is being dissipated.

Consider the previous analogy of the car driving back and forth in the driveway. Although it ended up where it began, energy (gasoline) was consumed, and work was performed (back-and-forth movement of the car). The term *root mean square* (rms) defines an ac voltage or current by comparison to an equivalent dc voltage or current. In other words, a 120-volt ac rms source is equivalent to a 120-volt dc source in its ability to dissipate power or perform work. (Values of rms are often called *effective values*.)

Consider the instantaneous point in time at the positive peak level illustrated in Fig. 9-3. This positive peak level is + 10 volts.

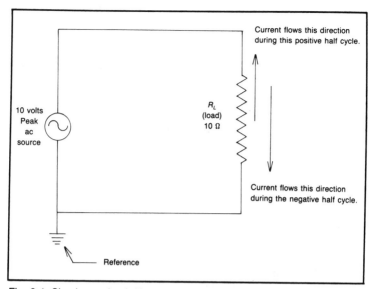

Fig. 9-4. Simple ac circuit illustrating current flow reversal.

If this potential is applied to a resistive load of 10 ohms, the instantaneous peak power dissipation can be calculated as follows:

$$P = \frac{E^2}{R} = \frac{(10)\,(10)}{10 \text{ ohms}} = \frac{100}{10} = 10 \text{ watts peak}$$

The preceding calculation proves that 10 watts of peak power is being dissipated during the positive peak voltage period. Other values of power are being dissipated at the other instantaneous voltage levels. At the instantaneous negative peak voltage, the same 10 watts of peak power is being dissipated. Thus, as long as current flows through a load, regardless of direction, power will be dissipated. For this reason, a term is needed to define the amount of work (power dissipation) an ac voltage is capable of performing as compared to a dc voltage. This is why rms is the most common way of defining the amplitude of ac voltages.

AC CALCULATIONS

NOTE: All of the following equations are applicable *only* to sine-wave ac voltages and currents.

If you know the peak or peak-to-peak value of a sine-wave ac voltage or current, you can calculate the rms value with Equations 9-3 or 9-4:

$$\text{rms} = (\text{peak}) \ .707 \qquad \textbf{(Equation 9-3)}$$

$$\text{rms} = (1/2 \ \text{peak-to-peak}) \ .707 \quad \textbf{(Equation 9-4)}$$

For example, use Equation 9-3 to calculate the rms value of the ac voltage shown in Fig. 9-3:

$$\text{rms} = (\text{peak}) \ .707 = (10) \ .707 = 7.07 \ \text{volts rms}$$

The ac voltage shown in Fig. 9-3 can be accurately defined as being 7.07 volts rms. This means it is equivalent to 7.07 volts dc. Currents of rms can be calculated in the same manner. If an rms voltage or current is known (sine wave only) and you wish to calculate the peak or peak-to-peak value, Equations 9-5 or 9-6 may be used:

$$\text{Peak} = (\text{rms}) \ 1.414 \quad \textbf{(Equation 9-5)}$$

$$\text{Peak-to-peak} = (2) \ \text{peak} \quad \textbf{(Equation 9-6)}$$

For example, common household outlet voltage is 120 Vac rms. You can calculate the peak and peak-to-peak values as follows:

$$\text{Peak} = (120) \ 1.414 = 169.68 \ \text{volts peak}$$
$$\text{Peak-to-peak} = (2) \ 169.68 = 339.36 \ \text{volts p-p}$$

You can use any of the ac voltage or current terms in the familiar Ohm's law equations. The only stipulation is that *they must be common*. In other words, if you use an rms voltage in an equation, you must use rms current, or the answer will be in error. For example, if you wanted to use Ohm's law ($E = IR$) to calculate a peak voltage, you must use the peak current. (Resistance is a constant value and, therefore, cannot be expressed in terms such as peak, peak-to-peak, or rms.)

For the following examples, refer to Fig. 9-4.

EXAMPLE 1: Calculate the rms voltage applied to the load.

$$\text{rms} = (\text{peak}) \ .707 = (10) \ .707 = 7.07 \ \text{volts rms}$$

EXAMPLE 2: Calculate the rms current flowing through the load.

$$I_{rms} = \frac{E_{rms}}{R} = \frac{7.07 \text{ volts}}{10 \text{ ohms}} = .707 \text{ amp rms}$$

EXAMPLE 3: Calculate the peak current flowing through the load.

$$I_{peak} = \frac{E_{peak}}{R} = \frac{10 \text{ volts}}{10 \text{ ohms}} = 1 \text{ amp peak}$$

EXAMPLE 4: Calculate the peak power being dissipated by the load.

$$P_{peak} = (I_{peak})(E_{peak}) = (1)(10) = 10 \text{ watts peak}$$

EXAMPLE 5: Calculate the rms (effective) power being delivered to the load.

$$P_{rms} = (I_{rms})(E_{rms}) = (.707)(7.07) = 5 \text{ watts rms}$$

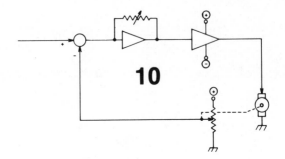

Inductance

THE MOST DIFFICULT ASPECT OF UNDERSTANDING INDUCTIVE devices is comprehending the basic concept. To begin with, inductors should always be thought of as storage devices. Water can be stored in a bucket; jelly, in a jar. Likewise, electrical energy can be stored in inductors.

At some point in your life, you may have built a small electromagnet by twisting some insulated electrical wire around a nail and connecting the ends of the wire to a flashlight battery, as shown in Fig. 10-1. This electromagnet is an *inductor*. An inductor is simply a coil of wire (not always coiled around a metallic core).

A typical inductor used in industrial electronics will consist of a coil (or multiple coils) wound on a metallic core. The metallic core concentrates the magnetic *flux lines* (lines of force) increasing the inductance value. Some coils used in high-frequency applications contain tunable slugs, which allow adjustment of the inductance value. Other types of high-frequency or special purpose inductors do not contain metallic cores. These are referred to as *air-core coils*.

The electromagnet shown in Fig. 10-1 illustrates how a coil will develop an electromagnetic field around it. The electromagnetic lines of force—the flux lines—flow through the nail, causing it to become a temporary magnet. Electromagnetic flux lines are stationary only as long as the current flow through the coil is constant. If the current flow through the coil is increased, the electromagnetic field will expand, causing the flux lines to move outward. If

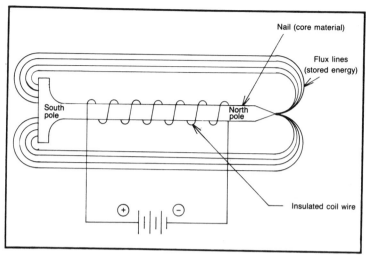

Fig. 10-1. Simple electromagnet.

the current flow is decreased, the electromagnetic field will collapse, causing the flux lines to move inward. The most important thing to understand is the field will move (expand and contract) as the current flow changes.

A very basic and important law of electricity follows:

Whenever an electrical conductor (wire) cuts magnetic flux lines, an electrical potential (voltage) will be produced in the conductor.

This is the basic principle behind the operation of any electrical generator. To generate electricity, magnetic flux lines must be cut by a conductor. This can be accomplished by either moving the conductor or moving the electromagnetic field.

As stated previously, if the current flow through a coil (inductor) is varied, the electromagnetic field will move in proportion to this variance. The movement of the electromagnetic field will cause flux lines to be cut by the conductor making up the coil. Consider the possible effects under the following three conditions:

1. The coil has a continuous and steady current flow through it; the applied voltage to the coil is held constant. Thus, the electromagnetic field surrounding the coil is also constant and stationary.
2. The applied voltage to the coil is reduced. The current flow

through the coil tries to decrease, causing the magnetic field surrounding the coil to collapse by some undetermined amount. As the field collapses, the flux lines cut the coil wire generating a voltage, which opposes the decrease in the applied voltage. The end result is that the coil tries to maintain the same current flow (for a period of time) by using the stored energy in the electromagnetic field to supplement the decreased current flow.

3. The applied voltage to the coil is increased. The current flow through the coil tries to increase, causing the magnetic field surrounding the coil to expand. As the field expansion occurs, the flux lines cutting the coil wire generate an opposing voltage. The end result is that the coil tries to maintain the same current flow (for a period of time) by storing energy in the electromagnetic field.

In essence, an inductor maintains a constant current flow by either expanding or contracting its associated electromagnetic field. The expansion and contraction of the electromagnetic field generates an opposing voltage referred to as the *counter-electromotive force* (cemf). Electromotive force is another term for voltage. The electromagnetic field represents stored energy. The quantity of energy that an inductor is capable of storing (inductance value) is measured in units called *henrys*. One henry is the inductance value of a coil if a current change of 1 ampere per second produces a cemf of 1 volt. The symbol for inductance is L.

Although this may seem somewhat complicated at first, consider the circuit in Fig. 10-2. (Note the symbol for a coil. The vertical lines to the right of the wire turns denote an iron core.) With the switch in the open position, obviously there cannot be any current flow through the inductor. The graph in Fig. 10-3 illustrates the current flow immediately after the switch in Fig. 10-2 is closed. The inductor tries to maintain the same current flow that existed prior to the closing of the switch (which was zero in this case) by storing energy in a magnetic field. While the magnetic field is expanding, a reverse voltage (cemf) is being generated by the coil opposing the applied voltage. The end result, as shown in Fig. 10-3, is that the current takes time to change even though the voltage is applied immediately.

The time period required for the current to reach its maximum value is measured in seconds and is defined by a unit called the *time constant*. The time constant is the time required (in seconds)

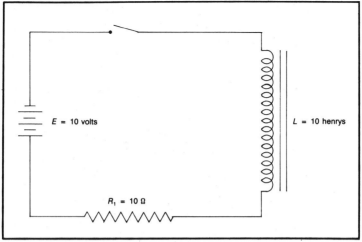

Fig. 10-2. Basic LR (inductive-resistive) circuit.

for the current flowing through the inductor to reach approximately 63 percent of its maximum value. The time constant is found by dividing the inductance value by the circuit resistance. For example, the time constant for the circuit shown in Fig. 10-2 would be:

$$T_c \ \frac{L}{R} = \frac{10 \text{ henries}}{10 \text{ ohms}} = 1 \text{ second}$$

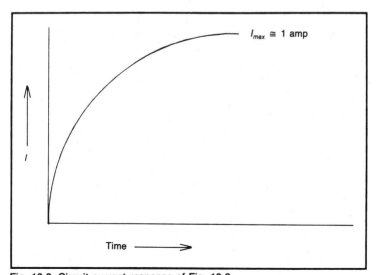

Fig. 10-3. Circuit current response of Fig. 10-2.

Assume that the maximum obtainable current in Fig. 10-2 is 1 amp. According to the previous time-constant calculation, the inductor current will climb to a value of approximately .63 amp after the switch has been closed for 1 second. During the next 1-second time period, the current will increase by another 63 percent of the remaining current. In other words, if the current rose to .63 amp after the first time constant, there would be .37 amp remaining before the maximum steady-state current was reached. During the second time constant, there would be a current increase of 63 percent of the remaining .37 amp; or approximately .23 amp. After two time constants, this results in an inductor current of approximately .86 amp. Likewise, during the third time constant, there would be another current increase of 63 percent of the remaining current. (This nonlinear increase is called an *exponential increase*.)

In theory, the maximum steady-state current can never be obtained. It would always be increasing by 63 percent of some negligible current value. In a practical sense, the maximum current value is usually assumed to have been reached after five time constants. The circuit shown in Fig. 10-2 would increase to its maximum 1 amp level in approximately 5 seconds.

The effect of the current trying to catch up with the applied voltage change is called the *current lag*. When the applied voltage is ac (changing constantly), the current will always lag behind the voltage. In a purely inductive circuit, this current lag is 90 degrees (regardless of the applied ac frequency). The circuit in Fig. 10-4 shows an inductor with an ac sine-wave applied voltage.

In Fig. 10-5, the top waveform shows the applied voltage. One cycle of any periodic waveform can be divided into 360 sections, or degrees. (This is the same manner in which points are defined on a circle. Any complete circle contains 360 degrees.) Figure 10-5 illustrates how a cycle is divided into 360 degrees (only the 90 increments are shown for the sake of clarity). Note the inductor's current waveform shown below it. When the applied voltage is at the 90-degree point (positive voltage peak), the current flow through the inductor is at the 0 degree point. When the voltage is at the 180-degree point (zero volts), the current is at the 90-degree point (positive peak). The current always lags behind the voltage by 90 degrees.

NOTE: The phase differential between voltage and current is commonly referred to as the *phase angle*. For example, if the voltage

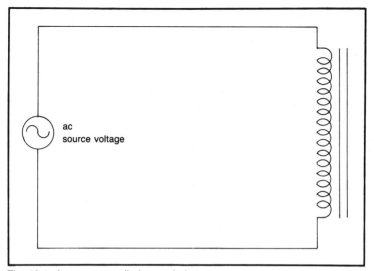

Fig. 10-4. Ac source applied to an inductor.

and current are out of phase by 30 degrees, the phase angle is said to be 30 degrees.

An interesting consideration of the circuit shown in Fig. 10-4 is the power consumption. Any purely inductive circuit does not dissipate power. An explanation of this phenomenon follows. Assume the peak current levels to be 10 amps and the peak voltage levels to be 10 volts. Referring to the phase differential shown in

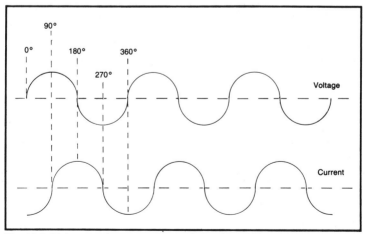

Fig. 10-5. Voltage and current phase relationship of Fig. 10-4.

87

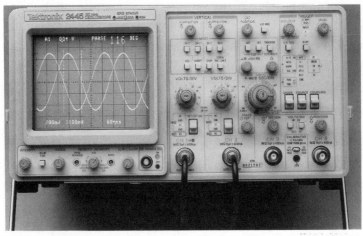

An oscilloscope displays a voltage and current phase differential of 116 degrees (courtesy of Tektronix, Inc.).

Fig. 10-5, you can calculate the instantaneous peak power dissipation at the 90, 180, and 270 degree points of the voltage waveform, as follows:

$$P = I\,E(\text{at 90 degrees}) = (0)\,(10) = 0$$
$$P = I\,E(\text{at 180 degrees}) = (10)\,(0) = 0$$
$$P = I\,E(\text{at 270 degrees}) = (0)\,(-10) = 0$$

Because of the 90-phase differential between voltage and current, the power dissipation in Fig. 10-4 is virtually zero. The inductor continually stores energy and then regenerates this energy back into the source.

If a voltmeter were used to measure the applied voltage of the circuit in Fig. 10-4, some value of rms voltage would be indicated. (If the applied voltage were 10 volts peak, the voltmeter would measure 7.07 volts.) Likewise, some value of rms current would be measured by an ammeter. If you were to multiply this voltage value by the current value, you would calculate some rms (or effective) power value. This contradicts the previous statement regarding zero power dissipation in a purely inductive circuit. For this reason, two new terms must be defined:

Apparent Power. The calculated power based on the rms voltage and current values.

True Power. The actual power dissipated when the voltage and current phase differential is taken into consideration.

In other words, when inductive circuits are involved in ac applications, you must always compensate for the voltage and current phase differential to calculate the true power dissipated.

NOTE: If you wish to pursue higher mathematics, true power is calculated by finding the apparent power and then multiplying it by the cosine of the differential phase angle. In Fig. 10-4, the differential phase angle is 90 degrees, and the cosine of 90 degrees is zero. Therefore, zero times any apparent power calculation will always equal zero.

Another consideration when dealing with inductance is the *power factor*. The power factor of any circuit is simply the true power divided by the apparent power. With the circuit shown in Fig. 10-4, the power factor would be:

$$\text{Power factor} = \frac{\text{true power}}{\text{apparent power}} = \frac{0}{(\text{any number})} = 0$$

Since the circuit shown in Fig. 10-4 is purely inductive, regardless of what the apparent power is, the power factor will always equal zero.

FREQUENCY CONSIDERATIONS

DC Resistance

Earlier in this chapter, the time constant of the circuit in Fig. 10-2 was examined, and the maximum obtainable current flow was found to be approximately 1 amp. You may be wondering why the resistive value of the inductor was not included in this calculation. Actually, the inductor will pose some dc opposition, but in many cases this resistance may be negligible. After five time constants, the current through the inductor will have reached its maximum value. From this point on, both the voltage and the current will remain stable (as long as the circuit is closed). Therefore, the electromagnetic field around the coil will also remain stable. If the electromagnetic field does not move, the wire making up the coil cannot cut flux lines. This means there cannot be any generation of cemf. In effect, the only opposition to current flow posed by the inductor is the wire resistance of the coil. The coil's wire resistance (usually called the dc resistance of the coil) is normally very low and may be disregarded in most applications.

The Q (abbreviation for quality) of an inductor is the inductance value (in henries) divided by the dc resistance (in ohms). For most inductors, this value will be 10 or higher.

AC Resistance (Inductive Reactance)

As the frequency of the applied voltage to an inductor is increased, the inductor's opposition to ac current flow increases. This is because the amount of energy capable of being stored in the inductor's electromagnetic field (inductance value) remains constant, but the time period of the applied ac voltage decreases. As the ac time period decreases, less energy is required from the inductor's electromagnetic field to oppose the voltage alternations. For example, it would take 10 times the energy to oppose 100 volts for 10 seconds than it would to oppose the same voltage for one second. The same principle applies with an increase in the frequency of the applied ac.

Consider the circuit shown in Fig. 10-4. As the frequency of the applied ac voltage is increased, the rms current flow in the circuit will decrease. This frequency-dependent opposition to current flow is called *inductive reactance*. If an ac ammeter is placed in this circuit, it will indicate a steady decrease in circuit current as the frequency of the applied voltage increases.

The current waveforms in Fig. 10-6 demonstrate this principle. The peak-to-peak current amplitude is much higher with the low frequency than it is with the high frequency (while the ampli-

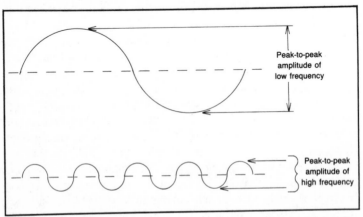

Fig. 10-6. Current amplitudes through an inductor at different frequencies whereby the source voltage constant is maintained.

tude of the applied voltage remains constant). Thus the inductive reactance increases as the frequency increases. Inductive reactance is defined in ohms.

NOTE: You should not confuse this change in opposition with a change in the phase difference between voltage and current. In a purely inductive circuit, the current will always lag behind the voltage by 90 degrees regardless of the frequency.

IMPEDANCE

When analyzing a circuit with inductance and resistance (such as that in Fig. 10-2), the total opposition to ac current flow will be a combination of the total resistance and the inductive reactance. (The total resistance is the resistance of R_1 plus the dc resistance of the coil.) This combined ac opposition is called *impedance*. Obviously, since the inductive reactance is frequency-dependent, the circuit impedance will also vary with frequency. Impedance is also defined in ohms. Its electrical symbol, as used in equations and formulas, is Z.

Unfortunately, the total circuit impedance is not calculated by simply adding resistance and reactance values. The mathematical procedures for calculating circuit impedances can become very complex. Fortunately, outside of highly complex electrical engineering, these calculations are not very useful. Since this book is not designed to train the reader to become an electrical engineer, this subject will not be covered.

TRANSFORMERS

An inductor with two (or more) coils wound in close proximity to each other is called a *transformer*. (A single coil inductor is often called a *choke*.) If an ac voltage is applied to one coil of a transformer, its associated moving magnetic field will cause the magnetic flux lines to be cut by itself (causing cemf) and any other coil near it. As the flux lines are cut by other coils nearby, an ac voltage is also induced in them. This is the basic principle of transformer operation.

If the multiple coils of a transformer are wound on a common iron core, the transfer of electrical energy through the moving electromagnetic field becomes very efficient. Iron-core transformers are designed for efficient transfer of power and are consequently called *power transformers* or *filament transformers*. Properly de-

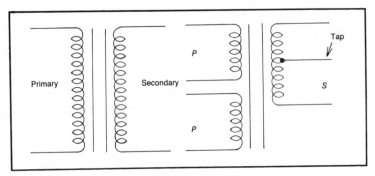

Fig. 10-7. Transformer symbols.

signed, these transformers can obtain efficiencies as high as 99 percent.

Figure 10-7 shows symbols for transformers. The coil on which the ac voltage is applied is called the *primary*. The transfer of power is induced into the *secondary*. Transformers may have multiple primaries and secondaries. The primaries and secondaries may also contain taps to provide multiple voltage outputs or to adapt the transformer to various input voltage amplitudes. The example shown on the left in Fig. 10-7 is a transformer with one primary and one secondary. The illustration on the right shows a transformer that has two primaries, one secondary, and a secondary tap. This transformer is *center-tapped* (ct)—that is, the tap is connected to the center of the winding.

The most important attribute of a transformer is its ability to increase or decrease ac voltage levels without significant loss of energy. This is accomplished by means of the *turns ratio* designed into the transformer. The turns ratio is simply the ratio of the number of turns on the primary to the number of turns on the secondary. For example, if a transformer has a 1:1 turns ratio, it has the same number of turns on the primary as it has on the secondary. A transformer with a 2:1 turns ratio has twice the number of turns on the primary as on the secondary. A transformer with a 1:12 turns ratio has twelve times the number of turns on the secondary as on the primary.

The turns ratio has a directly proportional relationship to the voltage level of the secondary. If the secondary has more turns than does the primary, the secondary voltage will be proportionally higher in amplitude, or *stepped up*. If the secondary contains less turns than does the primary, the secondary voltage will be proportionally lower, or *stepped down*. For example, if a transformer has a 2:1 turns ratio, as shown in Fig. 10-8, the secondary voltage read on the voltmeter will be one-half the amplitude of the primary. This is because there is only one-half the number of turns on the secondary as compared to the primary. The reverse is also true. If the transformer has a 1:2 turns ratio, the secondary voltage will be twice that of the primary.

A basic law of physics states that it is impossible to obtain more power from anything than is originally put into it. For this reason, if the secondary voltage is doubled, the secondary current will be halved. Likewise, if the secondary voltage is only half the primary voltage, the secondary current will be twice the value of the primary. In this way, a transformer will always maintain an equilibrium of power on both sides.

This principle is demonstrated in Fig. 10-9. Notice the transformer has a 10:1 turns ratio. This means whatever voltage is applied to the primary (E_p) will only be one-tenth that value on the secondary (E_s). The current flow is just the opposite. The current flowing through the load resistor (I_s) will be ten times greater than the primary current flow (I_p).

Assume that you apply 100 volts rms to the primary of the transformer in Fig. 10-9. Since there is only one-tenth the number of turns on the secondary, the secondary voltage will only be one-tenth the value of the primary, or 10 volts rms. The current flow through the load resistor will be 10 volts rms divided by 10 ohms, or 1 amp rms. Therefore, the power transferred to the secondary load will be:

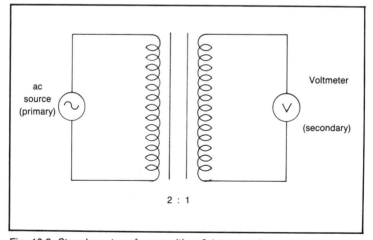

Fig. 10-8. Step-down transformer with a 2:1 turns ratio.

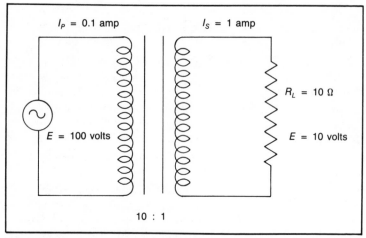

Fig. 10-9. Transformer with a secondary load demonstrating the relative primary and secondary currents.

$$P = IE = (1 \text{ amp}) (10 \text{ volts rms}) = 10 \text{ watts rms}$$

Now consider the power being delivered to the primary. The current in the primary will be one-tenth that of the secondary, or 0.1 amp rms. As previously stated, the applied voltage is 100 volts rms. Therefore:

$$P = IE = (0.1 \text{ amp rms}) (100 \text{ volts rms}) = 10 \text{ watts rms}$$

It is important for you to understand that the transformer itself is not dissipating 10 watts rms. It is transferring 10 watts rms of power from the primary to the secondary. The transformer itself is dissipating a negligible amount of power. If the secondary circuit were opened, the current flow in the secondary would consequently drop to zero. Likewise, the current flow in the primary would also virtually drop to zero. (A small current flow would still exist in the primary. This will be discussed later in this chapter.)

The effect of the secondary load on the primary current flow is explained by a phenomenon called *reflected impedance*. This simply means the impedance (or load) seen by the secondary will be reflected back to the primary. Referring back to the previous example problem, the primary current was calculated to be 0.1 ampere rms. Since you have been given the applied primary voltage (100 volts rms), you can use Ohm's law to calculate the primary impedance as follows:

$$Z = \frac{E}{I} = \frac{100 \text{ volts rms}}{0.1 \text{ amp rms}} = 1000 \text{ ohms}$$

You may have expected the primary impedance to be 10 times higher than the secondary load because of the 10:1 turns ratio. Actually, the reflected impedance presented by the primary is the square of the turns ratio times the secondary load. Therefore, the primary impedance of the circuit shown in Fig. 10-9 is:

$$Z_{primary} = (10 \times 10) \times 10 \text{ ohms} = 1000 \text{ ohms}$$

Notice that this answer agrees with the answer derived by Ohm's law. The same formula will work in reverse. For example, assume you know the primary impedance (1000 ohms) and you wish to calculate the secondary load. Since the secondary only contains 1/10 the number of turns as the primary and you are using primary values to calculate secondary values, you must consider the turns ratio as equal to 1/10, or 0.1. Therefore, the equation will be:

$$Z_{secondary\ load} = (0.1 \times 0.1) \times 1000 = 10 \text{ ohms}$$

If a transformer has multiple secondaries, the total power transfer of all of the secondaries will be reflected in the primary. The total reflected primary impedance can be found by calculating the individual reflected impedances from each secondary. These impedances are then considered to be in parallel. By solving for the equivalent parallel impedance, the primary impedance can be calculated.

For example, if the transformer shown in Fig. 10-9 had four identical secondaries with four identical loads, each secondary would reflect a 1000-ohm impedance to the primary (as explained previously). If you consider these four 1000-ohm impedances to be in parallel, you can calculate the equivalent parallel impedance just as you would calculate the total parallel resistance of four 1000-ohm resistors in parallel. The answer is 250 ohms. Therefore, the total reflected primary impedance would be 250 ohms.

Refer to Fig. 10-9 again. What would be the effect of removing the secondary load (resistor RL) from the circuit? The secondary voltage would still be 10 volts rms, but all secondary current flow would cease, because the circuit has been opened. Therefore, the secondary load impedance is infinite. Based on the equation for reflected impedance, this infinite load impedance is multiplied by the square of the turns ratio and reflected back to the primary. Obviously, infinity multiplied by any number is still infinity. This would mean the primary should present an infinite impedance to the 100 volt rms applied voltage. Actually, the primary impedance is not infinite; a small (or negligible) current flow will still remain in the primary even though the secondary load is infinite. There are two reasons for this:

1. Even in the best-designed transformers, small losses (called *eddy current* and *hysteresis* losses) will exist. In a typical power transformer, these losses may cause the transformer to become slightly warm without any load on the secondary circuit.

2. The maximum impedance that can be presented by the primary is the pure inductive reactance of the primary winding coil. In other words, when the secondary circuit of a transformer is opened, the secondary windings no longer affect the primary winding. Therefore, the primary winding looks like a simple coil, or choke. In a typical power (filament) transformer, the inductive value of this coil would be high (but not infinite). For example, a typical filament transformer primary may have an inductive reactance of 10,000 ohms at a 60-hertz frequency. If 120 volts rms are applied to this reactance value, a current flow of 12 milliamps will result. This 12-milliamp current flow would lag behind the applied voltage (120 volts rms) by approximately 90 degrees because the circuit is purely inductive. Therefore, the true power dissipation from this effect is zero.

Now, consider the opposite extreme. What happens to a transformer when the secondary is shorted (RL = 0 ohms)? Going back to the previous discussion on reflected impedance, the load resistance is multiplied by the square of the turns ratio. If the load resistance is zero, the answer will be zero regardless of the turns ratio. Again, there is a primary winding limitation. The primary impedance cannot actually drop to zero because of the dc resistance of the wire making up the primary coil. In a typical power transformer, this dc resistance will be very low (usually about 1 ohm or less). If 120 volts rms is applied to a typical transformer with a shorted secondary, it will try to dissipate about 14,000 watts ($P = E$ squared divided by R). Needless to say, it will not try to dissipate 14 kilowatts for very long before something "fries" (hopefully, a fuse)! With the shorted secondary, the primary should be purely resistive. Therefore, the primary current and voltage is in phase, and the apparent power equals the true power.

Consider the following points. If the secondary load is minimum (RL = infinity), the transformer primary is purely inductive, and the primary current lags behind the primary voltage by 90 degrees. Therefore, the power factor equals zero.

If the secondary load is maximum (RL = 0 ohms), the transformer primary is purely resistive, and the primary current and voltage are in phase. The power factor in this case is equal to one.

When the secondary load is somewhere between these two extremes, the phase differential between the primary voltage and current is also somewhere between 0 and 90 degrees.

The power factor of the primary changes proportionally to the secondary load. This is why a transformer will only transfer the needed power through the primary winding to the secondary load.

TRANSFORMER APPLICATIONS

Most modern solid-state equipment requires low-voltage power supplies to operate properly. Transformers are very useful in stepping down standard 120- or 240-volt ac power to some useful level.

A power transformer constructed with a 1:1 turns ratio can be used to electrically isolate one circuit from another. This type of transformer is referred to as an *isolation transformer*.

Transformers may also be used to match or tailor various impedances. They accomplish this by means of their reflected impedance characteristic.

A specially designed power transformer can be used to step-down 440/220 Vac to 120 Vac for the operation of control relays or contactors (big relays). This type of transformer is called a *control transformer*.

Power companies use large transformers to step-up voltages to very high amplitudes. This provides efficient power transfer over long distances with minimal-transmission cable diameter.

Constant-voltage, or *ferroresonant transformers* are commonly used for applications in which it is critical to maintain a constant secondary voltage even if the primary voltage fluctuates. These transformers have higher losses, causing a higher normal operating temperature than would conventional transformers.

INDUCTIVE CONSIDERATIONS

DC Circuits

Whenever inductors are used in electrical or electronic circuits, they involve stored energy. This stored energy can create some unexpected problems, which will be discussed in this section.

Consider the circuit shown in Fig. 10-10. When the switch SW 1 is closed, it will take approximately five time constants for the inductor current to reach its maximum value. During this time, a cemf will be generated, opposing the applied voltage. After the current reaches its maximum value, it will remain constant, because the applied dc voltage to the inductor is constant. This causes the electromagnetic field to become stationary, and all opposing cemf will cease. If the switch (SW1) is now opened, the inductor's current flow is abruptly eliminated. In a futile effort to try to maintain the same current flow, the electromagnetic field surrounding the inductor will quickly collapse. The inductor will actually try to cause a current to flow through an open circuit.

According to Ohm's law, in order to achieve a current flow through an infinite resistance (open switch), there must be an infinite applied voltage. If the inductor in Fig. 10-10 were perfect, it would collapse the electromagnetic field instantly and generate an infinite voltage.

Realistically, ordinary air does not pose an infinite resistance. Likewise, perfect inductors have not yet been designed. In reality, if the switch (SW1) in Fig. 10-10 is opened, an extremely high volt-

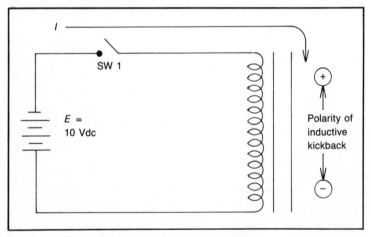

Fig. 10-10. Illustration of inductive kickback transients.

age spike (caused by the rapid collapse of the inductor's field) will appear across the open. This voltage spike could be thousands of volts in amplitude, but only for a very short duration. It is commonly referred to as the *inductive kickback voltage*.

NOTE: Sometimes, short duration events are termed *transients*. For this reason, inductive kickback voltages can be correctly referred to as *voltage transients*.

Inductive kickback voltages can cause numerous problems in electrical circuits. With the circuit shown in Fig. 10-10, the inductive kickback could arc across the switch contacts, causing the eventual failure of the switch. In electronic counting or control circuits, erroneous counts could be made or erroneous control actions instigated. Even worse, solid-state electronic devices can be destroyed by high-voltage spikes.

There is an easy method to eliminate inductive kickback transients. Notice the polarity of the inductive kickback voltage shown in Fig. 10-10. It will always be the opposite of the applied voltage, because it will try to maintain the same current flow in the same direction. If this is difficult for you to understand, consider the direction of the current flow when the switch (SW1) is closed. In Fig. 10-10, it flows from the top of the inductor to the bottom. When

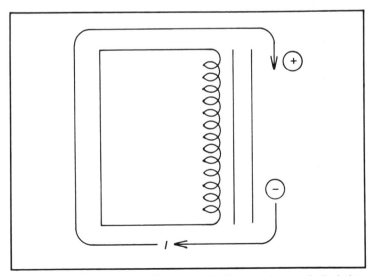

Fig. 10-11. Direction of current flow caused by the stored energy in the inductor illustrated in Fig. 10-10.

99

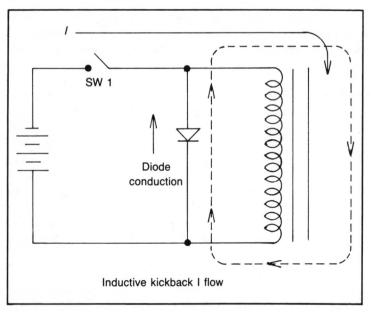

Fig. 10-12. Suppression of inductive kickback transients through the use of a diode.

the switch (SW1) is opened, the rapidly collapsing field causes the inductor to become a generator. Figure 10-11 demonstrates how this polarity would cause the current to flow in the same direction if the inductor faces a closed circuit. If a switch device existed that could sense the polarity reversal and short the inductor whenever an inductive kickback transient occurred, the transient could be eliminated.

A *diode* is such a device. Diode operation will be covered in a later chapter. For now, it is sufficient to say that a diode will conduct current in only one direction.

The circuit in Fig. 10-12 demonstrates this principle. When SW1 is closed, the normal direction of circuit current flow is opposite to the direction of the diode conduction. In this condition, the diode resembles an open and has no effect on the circuit operation. When SW1 is opened, the voltage reversal (due to the electromagnetic field collapse) causes the diode to resemble a short. Therefore, the electromagnetic field collapses at a rate sufficient to maintain the same current flow through the inductor until all of the energy in the electromagnetic field has been used up. In other words, the diode totally eliminates the inductive kickback transient but does not affect the normal circuit operation.

Electromechanical relays comprise the most common usage of inductors in dc circuits. The coil of a dc relay is an inductor. In virtually all applications requiring the use of a dc relay in conjunction with electronic circuits, a diode will be placed in parallel with the relay coil to eliminate the inductive kickback transient.

AC Circuits

Inductive kickback transients can also occur in ac applications, but the effects are usually not as severe. Usually, inductors used in ac applications are transformers. If any of the secondaries of a transformer remains connected to a load when the power to the primary is disconnected, the energy stored in the electromagnetic field is simply delivered to the load. This eliminates the high-voltage spike in all of the transformer windings, because they all share the same electromagnetic field. (Normally, the circuit power is interrupted on the primary side of a transformer while leaving the secondaries connected.)

When ac relays are used, inductive kickback transients will occur when the ac voltage to the relay coil is interrupted. In most applications, this does not cause any serious problems because the high-voltage spike can only cause a transient on the ac power lines. The ac power lines are usually isolated from any dc power supplies used for operating most electronic devices.

In applications in which it is desirable to suppress ac inductive kickback transients, devices such as capacitors and varistors may be incorporated. (These devices will be discussed in later chapters.)

TROUBLESHOOTING INDUCTORS

Specialized inductor testers are available from various manufacturers. Most of these specialized testers will provide excellent results along with accurate measurements of inductor parameters. Unfortunately, besides being very expensive, such equipment is not always available. The following methods are reliable and should provide good results when trying to isolate inductor problems.

Troubleshooting Coils

Only two possible problems can occur within a single coil inductor. The wire conductor making up the coil can break (or burn up), creating an open, or the individual windings can short, destroying the inductive characteristics of the coil.

101

It is very easy to troubleshoot a coil that has opened. Remembering that a coil poses very little dc opposition as compared to its ac opposition, you can simply measure the dc resistance with an ohmmeter. With very large inductors, you can expect to measure a virtual short (in milliohms). With small inductors, or *relay coils*, you may measure resistances up to several thousand ohms. The actual dc resistance is relatively unimportant. All you are trying to prove is the existence of continuity which indicates that the coil has not opened.

Trying to determine if the coil has shorted is much more difficult. I have used the following methods in the field. With a little practice and experience, you should be able to use these same methods with good success.

1. With small inductors, or small relay coils, the wire used to construct the windings is usually very small. This small wire presents a higher dc resistance than larger diameter wire. If the dc resistance of the coil is at least 10 ohms or higher, you can accurately measure it with an ohmmeter and use this reading in several ways:

   a. In many cases, the dc resistance is marked right on the coil. Simply measure the coil resistance with an ohmmeter, and compare the reading with the value marked on the coil.

   b. Comparison is a very important tool to use when specialized test equipment is not available. If a good spare coil of the same type is available, compare its resistance value with that of the coil in question. If the two measurements are within 10 percent of each other, the questionable coil is probably good.

   c. In some cases, the manufacturer of the coil or relay may be willing to provide information regarding the dc resistance of the coil over the telephone. Most reputable manufacturers are more than happy to help individuals use their products (especially if the manufacturer desires future orders from those individuals).

2. The following troubleshooting methods are applicable to virtually any coil:

   a. A good visual observation may detect a burned coil, which almost always indicates that the coil is defective. If the coil has overheated, it will usually show some discoloration. An overheated coil will also give off an odor (or even smoke if power is applied).

b. The conductor insulation in high-voltage coils may break down causing the coil windings to arc internally. If this happens, a frying or crackling sound will usually be present when power is applied. If this occurs, the coil should be replaced.

c. If none of the above methods is applicable, and test equipment such as rcl bridges or Z meters are not available, the troubleshooter must rely on his experience and make an educated guess. If the coil wire is very small and there are many turns on the coil, the troubleshooter can expect to measure a significant resistance and vise versa.

Troubleshooting Transformers

As in the case of single-coil inductors, each winding of a transformer should be tested for continuity. Any open windings indicate that the transformer is defective. There should not be any dc continuity between individual windings of a transformer. In addition, there should not be any continuity from any of the windings to the core material.

If a single transformer wire does not show continuity to anything (except possibly the core), the shield wire is probably the culprit. Some transformers are manufactured with a foil shield around the coil windings to reduce the effect of undesirable electrical noise (the shield may also be connected to the core material).

Testing a power transformer for internal shorts can be much easier (and more definite) than single-coil inductors. The following method should give you a good indication of the transformer's condition.

NOTE: Only qualified electrical personnel familiar with high-voltage safety procedures should attempt the following tests.

It is much more common for a transformer to burn-up than it is to open. If a transformer develops too much internal heat, the insulating varnish on the coil windings melts, causing the windings to short together. This condition will usually result in a blown primary-winding fuse. A blown primary-winding fuse will also result if the load on the transformer's secondary winding (or windings) is excessive. If you discover a blown primary-winding fuse, you must determine which of these conditions exist (or if they both exist). The easiest way to make this determination is as follows:

1. Turn off all power to the transformer and associated circuits.
2. Disconnect all of the secondary windings from their associated loads.
3. Re-apply power to the transformer primary. If the primary fuse blows again, the transformer is defective and should be replaced.
4. If the primary fuse doesn't blow in Step C, continue to apply power to the transformer primary for at least one-half hour. After one-half hour, check the temperature of the transformer. If it is in excess of approximately 150 °F, the transformer probably has one or more shorted windings and should be replaced. If the transformer is cool, a fault exists in one (or more) of the secondary loads.

Shorted transformer secondaries will always cause excessive current flow in the transformer primary (due to the reflected impedance) which will either blow the primary fuse or cause excessive heat buildup in the transformer. Any transformer can be checked on the bench with the following technique: simply apply primary power, and wait for the transformer to heat up. (While waiting for transformer heat buildup, measure and verify the secondary voltages.) When connecting the transformer primary to an ac source, use extreme caution to observe any primary voltage taps (if applicable), and make the proper connections as specified by the manufacturer.

Be sure to allow adequate time for the transformer to heat up (approximately 30 minutes). If it becomes painful to hold a finger on the transformer housing for more than one second, it is probably above 150 °F. (Disconnect power to the transformer before touching it.) If a transformer becomes this warm without any secondary load on it, it is probably defective and should be replaced.

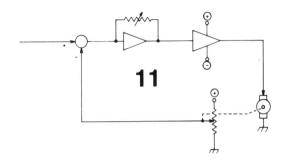

11

Capacitance

A *CAPACITOR* IS ONE OF THE MOST COMMON COMPONENTS used in electronics but is probably one of the least understood. As in the case of inductors, a capacitor is a storage device. An inductor, as you learned in Chapter 10, stores electrical energy in the form of an electromagnetic field, which collapses or expands to try to maintain a constant current flow through the coil. Similarly, a capacitor stores a static charge, which increases or decreases to try to maintain a constant voltage across the capacitor.

CAPACITOR CONSTRUCTION

A capacitor consists of two conductive plates. Between the plates is an insulator called the *dielectric.* The size of the plates and the thickness of the dielectric determine the capacity (storage capability). The capacity may be increased through the use of larger conductive plates or thinner dielectric material. The thinner the dielectric, the higher the capacity, but the lower the maximum voltage rating. Because it is impractical to manufacture (or try to use) extremely large metal plates, capacitors are usually manufactured by rolling the plates between the dielectric in a round tubular form resembling a firecracker.

CAPACITOR PRINCIPLES

NOTE: The following theory of physics involved in the operation of a capacitor has long been accepted as correct. In later years, a different theory of physics involved with capacitor operation has emerged. The old theory is presented here because it will help you understand how a capacitor works. The more recent theory will be explained at the end of this chapter.

When an electrical potential (voltage) is placed across the two plates of a capacitor, a certain number of electrons are drawn from one plate and flow into the positive terminal of the source. At the same time, the same number of electrons flow out of the negative terminal of the source and are pushed into the other plate of the capacitor. This process continues until the capacitor is charged to the same potential as the source. When this occurs, all circuit current flow ceases. The dielectric (insulator) material does not allow current to flow from plate to plate.

Figure 11-1 shows a capacitor in series with a resistor and an applied source voltage of 10 Vdc. When the switch (SW1) is first closed, a circuit current will begin to flow. The rate of the current flow (which determines how quickly the capacitor will fully charge) will be limited by the series resistance in the circuit and the difference in potential between the capacitor and the source.

As the capacitor begins to charge (build up electrical pressure), the rate of current flow from the source to the capacitor begins to decrease. The graph shown in Fig. 11-2 demonstrates the voltage drop across the capacitor (C_1) relative to time. The capacitor eventually charges to the full potential of the source voltage (10 volts). When this happens, all circuit current will cease. If the switch is

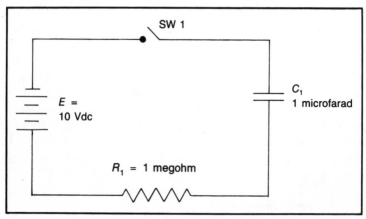

Fig. 11-1. Basic RC (resistive-capacitive) circuit.

opened at this time, the capacitor will hold this static charge until given a discharge path of lesser potential.

The capacitor eventually charges to a maximum electrical pressure of 10 volts. The capacity, or quantity of charge it can hold at this potential, is determined by the physical characteristics of the capacitor.

If you are confused at this point, consider the following analogy. Assume that an air compressor is adjusted to maintain 10 pounds per square inch (psi) of air pressure in the compressor's holding tank (this is analogous to the 10-Vdc source in Fig. 11-1). From this holding tank, you wish to pressurize a small portable air tank to the same 10 psi pressure (the small portable air tank is analogous to the capacitor).

As soon as you connect an air hose from the holding tank to the small portable tank, air begins to flow (just as current flows) rapidly from the holding tank to the small tank. The rate of this air flow (which determines how fast the small tank will fully pressurize) will be limited by the size of the air hose connecting the tanks and the difference in air pressure between the tanks.

As the small tank begins to pressurize, the rate of air flow from tank to tank begins to slow down. This is because the pressure differential is decreasing. Eventually, the small tank is fully pressurized to the same pressure as the holding tank. When this happens, all air flow from tank to tank ceases. If you disconnect the small tank at this time, it will hold this pressurized air (at 10 psi) until you open its release valve (which is analogous to providing

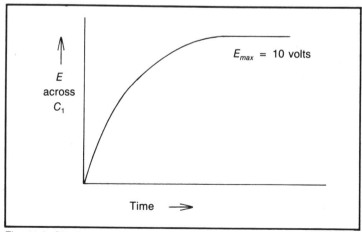

Fig. 11-2. Circuit voltage response of Fig. 11-1.

a discharge path of lesser potential).

As stated previously, the final air pressure obtained in the small tank is 10 psi. The capacity, or quantity of air it can hold at this pressure, is determined by the physical size of the air tank (which is analogous to the quantity of charge a capacitor is capable of holding).

The capacity (quantity of charge) of a capacitor is measured in units called *farads*. A farad is the amount of capacitance required to store one coulomb of electrical energy at a 1 volt potential. A *coulomb* is a volume measurement unit of electrical energy (charge). It is analogous to volume measurement units, such as quart, pint, or gallon. A gallon represents four quarts. A coulomb represents 6.28×10^{18} electrons (or 6,280,000,000,000,000,000 electrons).

The basic unit of current flow, the ampere, can be defined in terms of coulombs. If one coulomb passes through a conductor in one second, this is defined as one ampere of current flow. Thus 1 ampere = 1 coulomb/sec.

A farad is generally too large a quantity of energy to be stored by only one capacitor. Therefore, the capacity of most capacitors is defined in terms of *microfarads* or *picofarads*. As stated previously, capacitors also have an associated voltage rating. If this voltage rating is exceeded, the capacitor could develop an internal short.

As the graph in Fig. 11-2 shows, a capacitor charges exponentially. This exponential voltage change across a capacitor is identical to the exponential current change through the inductor shown in Fig. 10-3 of Chapter 10. Likewise, the time required for voltage to reach its maximum value is measured in seconds and is defined by the *time constant.* The time constant is equal to the amount of time required for the voltage across the capacitor to reach a value of approximately 63 percent of the maximum applied voltage. To calculate the time constant, simply multiply the capacitance (in farads) times the resistance (in ohms). For example, the circuit in Fig. 11-1 would have a time constant of:

$$T_c = C\,R = (.000001)\,(1,000,000) = 1 \text{ second}$$

This problem would be much easier to calculate using scientific notation:

$$T_c = C\,R = (1 \times 10^{-6})\,(1 \times 10^6)$$

The negative sixth power cancels out the positive sixth power, leaving this simplified problem:

$$(1)(1) = 1 \text{ second}$$

The principle of the time constant that applies to inductance also applies to capacitance. During the first time constant, the capacitor charges to approximately 63 percent of the applied voltage. The capacitor in Fig. 11-1 would charge to approximately 6.3 volts in 1 second after SW1 had been closed. During the next time constant, the capacitor would charge to an additional value of 63 percent of the remaining voltage differential. Five time constants are required for the voltage across the capacitor to obtain the value usually considered to be the same as the source voltage. (Remember, in theory, the maximum applied voltage can never be obtained, because it would always be increasing by 63 percent of some remaining voltage value.)

In the circuit shown in Fig. 11-1, the approximate source voltage (10 volts) would be obtained across the capacitor in 5 seconds. If the switch (SW1) was opened (after being previously closed for at least 5 seconds) the capacitor would hold this stored energy at a 10-volt potential indefinitely.

Consider the current and voltage relationship in Fig. 11-1 when the switch is closed (assuming the capacitor has been discharged). Immediately after switch SW1 is closed, the capacitor offers virtually no opposition to current flow, because it is discharged (just like the non-pressurized air tank in the earlier analogy). This maximizes the current flow and minimizes the capacitor's voltage. As the capacitor begins to charge, the current flow begins to decrease. At the same time, the voltage across the capacitor begins to increase. When the capacitor is fully charged, the current flow will cease (just as the air flow ceased in the earlier analogy when the small air tank reached the same pressure as the air compressor's holding tank). The voltage across the capacitor will equal the source voltage. The important point for you to remember is that the peak current occurs before the peak voltage is developed across the capacitor. Simply stated, the voltage lags behind the current in a capacitor.

A capacitor has the unique ability of trying to maintain a constant voltage. The circuit in Fig. 11-3 demonstrates this principle. When the switch SW1 is closed, capacitor C_1 begins to charge. It will charge at a rate based on the time constant of the circuit. In this case, there is no series resistance (except for the negligible re-

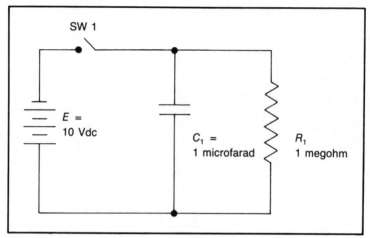

Fig. 11-3. Circuit used to demonstrate the C_1 charge time constant versus the discharge time constant.

sistance of the circuit wire and the internal resistance of the source). Therefore, the time constant will be:

$$T_c = R\,C = (0)\,(1 \times 10^{-6}) = 0$$

Thus, the capacitor will almost instantly charge to the 10-volt value of the source. Once the capacitor is charged, the current flow through the capacitor circuit will cease. But current will continue to flow through the parallel resistor circuit (consisting of R_1). The current flow through R_1 will be:

$$I = \frac{E}{R} = \frac{10}{1 \times 10^6} = 10 \text{ microamps}$$

At this point, if the switch (SW1) is opened, all current flow from the battery will cease. But there will still be current flow through the resistor R_1. This current flow comes from the stored energy in the capacitor C_1. In other words, C_1 becomes the temporary source of current flow through R_1 until it has exponentially discharged.

The discharge time constant can be calculated in the same manner as the charge time constant. In this circuit, the discharge time constant will not be the same as the charge time constant because, during the capacitor's charge, the series resistance from the source is negligible. During the discharge (when SW1 is opened), C_1 must

110

discharge through the 1-megohm resistance of R_1. Therefore, the discharge time constant will be:

$$T_c = R\,C = (1 \times 10^6)\,(1 \times 10^{-6}) = 1 \text{ sec.}$$

This means the voltage across C_1 (which will be the same voltage across R_1, because C_1 and R_1 are in parallel) will drop 63 percent in one second after SW1 is opened. In other words, approximately 3.7 volts will remain across R_1. During the second discharge time constant, the voltage across R_1 will decrease by 63 percent of 3.7, and so forth. The capacitor C_1 can be considered to be fully discharged after five time constants.

If capacitor C_1 had not been in this circuit, the voltage across R_1 would have disappeared immediately after SW1 was opened. When C_1 is placed in the circuit, it tries to maintain the same voltage even after the source is removed. To take this example one step farther, assume that SW1 is periodically opened and closed every 0.1 second. If C_1 is not in the circuit, the voltage across R_1 (and the current flow through R_1) would appear and disappear every 0.1 second. With C_1 in the circuit, the voltage remains relatively constant across C_1 even though the switch (SW1) is opening and closing every 0.1 second. C_1 is able to take on the charge very rapidly due to the very rapid charge time constant, but it cannot discharge rapidly due to the 1-second discharge time constant. C_1 will maintain the voltage across R_1 relatively constant as long as SW1 opens and closes at a periodic rate which is significantly faster than the discharge time constant.

If this is difficult for you to understand, consider this analogy. Assume you are trying to inflate a punctured bicycle tire with a hand pump. The air pressure applied by a typical hand pump is not continuous. It exists only during the down stroke. Hence, the air from the bicycle pump is actually delivered in pulses, or bursts of air. The pulsating air from the hand pump is analogous to the opening and closing of SW1. Because the bicycle tire has a hole in it, it will lose air as you attempt to fill it. If the hole is small, the hissing sound from the escaping air will be continuous even though you are supplying air in pulses. As long as you pump faster than the air can escape, the tire pressure (and the escaping air flow) will remain relatively constant. This is analogous to the voltage across R_1 (and the current through R_1) remaining relatively constant even though the applied source voltage is pulsed.

It is important to realize that as a certain amount of power (volt-

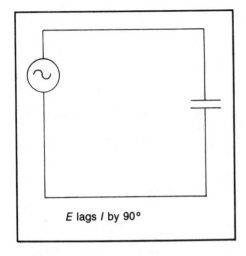

Fig. 11-4. Ac source applied to a capacitor.

E lags *I* by 90°

age times current) flows into a capacitor, the same amount flows out of it. The capacitor simply smooths the peaks and valleys into a continuous flow.

Figure 11-4 shows a circuit with a capacitor and an applied ac voltage. As stated previously, the voltage lags behind the current in capacitive circuits. This voltage lag is 90 degrees if the circuit is purely capacitive, as in Fig. 11-4. As in the case of inductance, if the phase differential between voltage and current is 90 degrees, the power factor is zero, and there is no true power dissipation. The circuit in Fig. 11-4 would not dissipate any power, because the voltage will always lag behind the current by 90 degrees regardless of the frequency.

CAPACITIVE REACTANCE

Figure 11-5 shows a series resistive and capacitive circuit with an applied ac voltage. As stated previously, a capacitor blocks dc current after it charges to the full applied dc potential. In an ac application, such as Fig. 11-5, there will be a continuous ac current, because the capacitor is continually charging and discharging as the applied voltage polarity is continually reversing.

In a dc application, when the capacitor begins to charge, the circuit current begins to drop. In other words, the maximum current occurs when the capacitor first begins to charge. After several time constants, the current flow is drastically reduced from the initial value. This nonlinear current characteristic has a profound effect when an ac voltage is applied to a capacitor. If the frequency

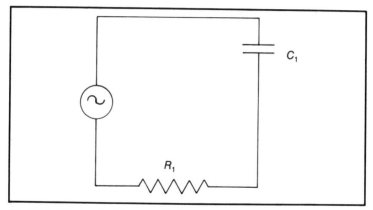

Fig. 11-5. RC circuit with an applied ac source.

of the applied ac is high enough to continually reverse during the early part of the capacitor charge period, there will be a higher rms current flow in the circuit. If the frequency of the applied ac is relatively low, the continuous reversal will occur during a later part of the charge period, causing the rms current flow to decrease. For example, as the applied ac frequency is increased to the circuit shown in Fig. 11-5, the rms current will also increase and vice versa. This means the opposition to ac current flow presented by the capacitor decreases as the frequency of the applied voltage increases.

This frequency-dependent opposition to current flow is called *capacitive reactance*. Like any type of opposition to current flow, capacitive reactance is defined in ohms.

Technically speaking, it is incorrect to state that ac current flows through a capacitor; the dielectric material (insulator) should not allow electrons to flow through it. Actually, a capacitor simply stores a charge that is released when the polarity is reversed. However, the continual charge and discharge (with an applied ac) gives the effect of ac current passing through the capacitor. Therefore, it is acceptable to define this effect in terms of frequency-dependent opposition (capacitive reactance).

This circuit in Fig. 11-5 contains both capacitive reactance and resistance. As in the case of inductance, the total opposition to current flow in the circuit is called *impedance*. Impedance is also defined in ohms.

CAPACITOR LEAKAGE

Some capacitors, especially those of the electrolytic variety, allow a very small dc current to flow through the dielectric material. This small current is called *leakage*. Leakage is an undesirable characteristic and is reduced as much as possible during the capacitor's manufacturing process.

CAPACITIVE AND INDUCTIVE COMPARISON

The following table shows the characteristics and inverse similarities between inductance and capacitance. Take a few moments now to study these relationships.

INDUCTANCE	CAPACITANCE
Voltage leads the current	Voltage lags behind the current
Tries to maintain a constant current.	Tries to maintain a constant voltage.
$$T_c = \frac{L}{R}$$	$$T_c = R\,C$$
As frequency increases, reactance increases.	As frequency increases, reactance decreases.
Ac power dissipation is zero in a purely inductive circuit.	Ac power dissipation is zero in a purely capacitive circuit.

CAPACITOR TYPES

Basically, all capacitors can be divided into two categories—*polarized* and *nonpolarized*, as shown in Fig. 11-6. If a capacitor is polarized, the correct polarity must be maintained when using it. A polarized capacitor will indicate, through labels on its body, which lead is connected to the positive polarity and which lead is connected to the negative polarity. Accidental reversal of this polarity can cause a polarized capacitor to explode. Invariably, the capacitor will be ruined.

With nonpolarized capacitors, the lead polarity is not critical.

PAPER AND MICA

Paper and mica were the standard dielectric materials for many years. Nowadays, many new dielectrics with various ratings are used in capacitors. Although mica is very seldom used in modern products, paper is still used quite often. The paper is impregnated

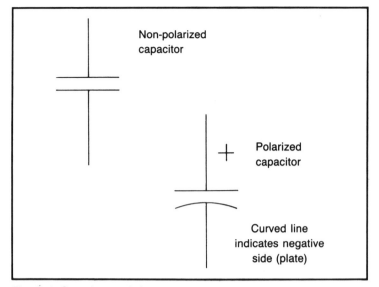

Non-polarized
capacitor

Polarized
capacitor

Curved line
indicates negative
side (plate)

Fig. 11-6. Capacitor symbols.

with a wax or special oil to reduce air pockets and moisture absorption.

PLASTIC FILMS

Plastic films of polyester, polycarbonate, polystyrene, polypropylene, and polysulfone are used in many of the newer large-capacity, small-size capacitors. Each film has its own special feature and is chosen to be used in a circuit for this feature. Some of the plastic films are also vacuum-plated with a metal. These are generally called *self-healing* capacitors and should not be replaced or substituted with any other type of capacitor.

CERAMIC

Ceramic is the most versatile of all dielectric materials. Many variations of capacity can be created by altering ceramic material. Capacitors that increase, stay the same value, or decrease value with temperature changes can be made.

If a ceramic disc is marked with a letter P such as P100, then the value of the capacitor will increase 100 parts per million per degree centigrade increase in temperature. If the capacitor is marked NPO or COG, then the value of capacity will remain constant with an increase in the temperature. Ceramic disc capacitors marked with an N, such as N1500, will decrease in capacity as the temperature increases.

Reaction to temperature changes is called *temperature coefficient*. This term simply refers to the manner in which a component is affected by changes in temperature. If a component has a negative temperature coefficient, its value decreases as the temperature increases and vice versa. If it has a positive temperature coefficient, its value increases as the temperature increases and vice versa. A capacitor's temperature coefficient is critical for circuits in which minor changes in capacitance can adversely affect the proper operation of the circuit.

A ceramic capacitor marked GMV means the marked value on the capacitor is the *guaranteed minimum value* of capacity at room temperature. The actual value of the capacitor can be much higher. This type of capacitor is used for applications in which the actual value of capacity is not critical.

Ceramic capacitors have been the most commonly used capacitors in electronics because of the versatility of the different temperature coefficients and the low cost. When replacing a ceramic disc capacitor, be sure to replace the defective capacitor with one having the same characteristics and voltage rating.

ELECTROLYTICS

Aluminum electrolytic capacitors are very popular because they provide for a large value of capacity in a small space. Electrolytic capacitors are polarized, and the correct polarity must always be observed when using or replacing these devices. For special-purpose applications, nonpolarized electrolytics can be obtained.

The aluminum electrolytic capacitor is constructed with pure aluminum foil wound with a paper soaked in a liquid electrolyte. When a voltage is applied to this combination, a thin layer of oxide film forms on the pure aluminum. This film becomes the dielectric; the oxide acts as a good insulator. As long as the electrolyte remains liquid, the capacitor is good or can be reformed by applying a dc voltage to it for a period of time (observing the correct polarity). When the electrolyte dries out, the leakage increases, and the capacitor loses capacity. This is called *dielectric absorption*.

Dielectric absorption can happen to aluminum electrolytics even in storage. Sometimes, if an electrolytic capacitor has been "sitting on the shelf" for a long period of time, it may need to be reformed to build up the oxidation layer. This can be easily accomplished by connecting it to a dc power supply for approximately an hour. *Remember to observe the correct polarity, or the capacitor may explode.* Most electrolytic capacitor manufacturers do not guarantee a capacitor shelf-life for more than five years.

TANTALUM ELECTROLYTICS

Although aluminum electrolytics are still the most popular type, tantalum electrolytics are also gaining popularity. Tantalum capacitors have several advantages over aluminum electrolytics—lower leakage, tighter tolerances, and small size for an equivalent capacity. Unfortunately, their size is limited and for large-capacity values, aluminum electrolytics are still preferred.

TESTING CAPACITORS

Various faults can occur within a capacitor. It can internally open, short, develop excessive leakage, break down under normal voltage levels, or change value. The only sure way to properly test a capacitor is with a specially designed dynamic capacitor tester. (An impedance bridge will not adequately test a capacitor, because it will not apply the normal operating voltage to test for dynamic breakdown or internal arcing.)

A typical capacitance meter (courtesy of B & K-Precision/Dynascan Corp.).

Fortunately, with the exception of electrolytic and tantalum capacitors, the failure rate is extremely low in other types of capacitors (unless some form of physical damage to the capacitor is detected). For occasional servicing requirements, an expensive dynamic tester may not be cost-effective. If a dynamic tester is not available, you can use the following procedures to be reasonably sure of the condition of a questionable capacitor.

Paper, mica, plastic, and ceramic capacitors are probably good if they do not read shorted when tested with an ohmmeter. Be sure the ohmmeter is not reading some other component in parallel with the capacitor in question. A simple way to prove this is to temporarily disconnect one side of the capacitor before taking the reading.

When testing electrolytic capacitors with an ohmmeter, you must first consider one aspect of the operation of the ohmmeter. An ohmmeter actually measures resistance by applying a small voltage to the unknown resistance and then calculates the resistance by measuring the current flow through it. A typical ohmmeter applies approximately 1.5 volts to the unknown resistance. This small voltage can be used to your advantage in making a simple test to the capacitor. First, disconnect one side of the capacitor in question to eliminate any parallel circuit resistance. Set the ohmmeter to a low scale. Connect the positive (red) lead of the ohmmeter to the positive side of the capacitor and the negative (black) lead to the negative side. Allow a few seconds for the capacitor to charge to the voltage applied by the ohmmeter, and then disconnect the ohmmeter from the capacitor. Switch the ohmmeter to measure voltage on the closest scale to 1.5 volts. Finally, measure the charge on the capacitor. If the capacitor is good, the voltmeter should read approximately the same voltage that was applied to the capacitor by the ohmmeter (about 1.5 volts). This voltage will slowly bleed off at a rate depending on the size of the capacitor and the impedance of the voltmeter being used to measure it.

The important point of this test is to determine whether the capacitor in question is capable of holding a charge. If it does not hold the small charge given it by the ohmmeter, it is probably bad. If it does hold the charge, you can assume it is good. However, you cannot determine whether it will break down under a higher voltage or if its rated capacitance value is correct.

PHYSICS PRINCIPLES OF CAPACITANCE

As stated in the beginning of this chapter, the physical principle behind the operation of a capacitor would be discussed briefly. This is not necessary information, but you might find it interesting.

Michael Faraday's theory more closely approaches the way a capacitor really works. He stated that the charge is in the dielectric material—not on the plates of the capacitor. Inside the capacitor's dielectric material, there are tiny electric dipoles. When a voltage is applied to the plates of the capacitor, the dipoles are stressed and forced to line up in rows, creating stored energy in the dielectric.

By now, the dielectric has undergone a physical change similar to that of soft iron when it is exposed to magnetic flux lines and becomes a magnet. If you removed the dielectric of a charged capacitor and then measured the voltage on the plates of the capacitor, you would discover the voltage was no longer present. By reinserting the dielectric and measuring the plates, you would find the voltage to which the capacitor had been charged before the dielectric had been removed. Thus, the charge of the capacitor is actually stored in the dielectric material. When the capacitor is discharged, the electric dipoles become reoriented at random, discharging their stored energy.

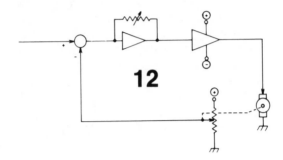

12

Resonance

YOU HAVE LEARNED THAT CAPACITORS AND INDUCTORS ARE frequency-dependent. In other words, their reactance (capacitive or inductive) depends upon the frequency applied to them. You should recall that capacitive reactance decreases as the frequency increases, while inductive reactance increases as the applied frequency increases. In simple terms, their reaction to frequency changes are exactly opposite. This leads to some interesting effects when inductors and capacitors are combined in electronic circuits.

Consider the series-resonant circuit shown in Fig. 12-1. The source is a variable-frequency ac source, which allows the user to vary the applied frequency throughout a wide range. (A source of this type is commonly used for testing purposes and is called a *signal generator* or *function generator*.) This variable frequency is applied to a series circuit containing some value of inductance and capacitance. Regardless of the ac frequency applied, both the inductor and capacitor will pose some value of reactance.

If the applied ac source is of very low frequency, the inductor will present very little opposition, but the capacitive reactance will be very high. On the other hand, if the ac frequency is increased to a very high value, the capacitor will now present very little opposition, but the inductive reactance will be very high. In either of these two cases, very little circuit current will flow in the circuit, because either one or the other reactive devices will pose a very high reactance.

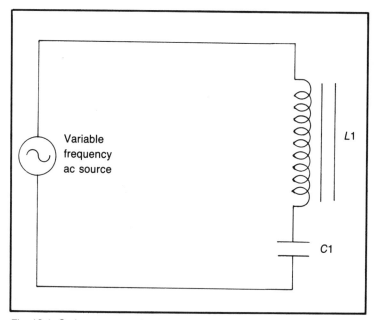

Fig. 12-1. Series-resonant circuit.

As the frequency of the variable ac source is adjusted through-out its entire range, a specific frequency will be reached that will cause the inductive reactance to equal the capacitive reactance. At this specific point in the frequency spectrum, the circuit current will be the highest, because it is not being blocked by either reactive component. From a mathematical viewpoint, the inverse characteristics of inductive reactance and capacitive reactance cancel one another, causing only the resistive element to remain in the circuit. This is called the *point of resonance*. Resonance is always obtained when the capacitive reactance is equal to the inductive reactance.

Figure 12-2 illustrates the effect upon circuit current as the frequency of the applied ac is varied in the circuit shown in Fig. 12-1. Notice how the circuit current peaks at a specific frequency but is substantially reduced at other frequencies above or below this point. The specific frequency causing the reactance values to equal each other is called the *resonant frequency*. The formula for calculating the resonant frequency is shown in Equation 12-1.

$$f_r = \frac{1}{6.28 \sqrt{LC}} \qquad \textbf{Equation 12-1}$$

where f_r is the resonant frequency in hertz;
             L is the inductance in henrys;
             C is the capacitance in farads.

When a capacitor and an inductor are placed in series, the circuit current will always be highest at the point of resonance. Thus, the circuit impedance at resonance is at its lowest point.

Consider the parallel-resonant circuit shown in Fig. 12-3. As can be seen in this circuit, the inductor and capacitor are now in parallel. Once again, the source is a variable-frequency ac source. Assume that the variable frequency ac source is adjusted to a low frequency. The source current will be high, because the inductor will offer very little inductive reactance to the low frequency. This will allow the source current to freely pass through the inductive leg of the parallel circuit. If ac source is adjusted to a high frequency, there will still be a high source current, because the capacitor will offer very little capacitive reactance to the high frequency, and the source current will freely pass through this parallel leg. In either of these two cases, the source current will al-

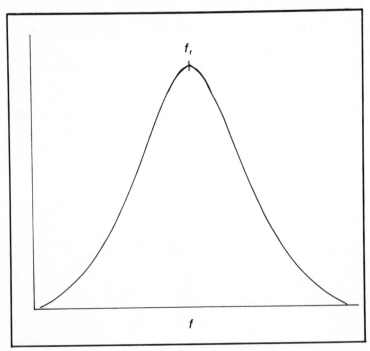

Fig. 12-2. Current versus frequency response of Fig. 12-1.

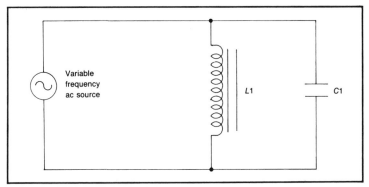

Fig. 12-3. Parallel-resonant circuit.

ways be high, because one or the other reactive legs will always have a low reactance.

As the ac source is varied over its entire range, a specific frequency will be reached that will cause the inductive reactance to equal the capacitive reactance. At this frequency, the overall source current will be at its minimum. This is the resonant frequency for the parallel circuit. Any frequency above or below this point will cause the source current to increase.

Figure 12-4 illustrates the effect on source current as the frequency is varied throughout its range. Note that the source current is at its lowest point at the resonant frequency but climbs on either side of this frequency. In a parallel-resonant circuit, the circuit impedance is at its highest at the point of resonance.

If you need to calculate a resonant frequency, Equation 12-1 is applicable for either series or parallel resonant circuits. In addition, Equations 12-2 and 12-3 allow you to calculate capacitive or inductive values if the desired resonant frequency is known.

$$L = \frac{1}{4\pi^2 f^2 C} \qquad \textbf{Equation 12-2}$$

where C is the capacitance in farads;
         π is 3.14 (Pi);
         f is the frequency in hertz;
         L is the inductance in henrys.

$$C = \frac{1}{4\pi^2 f^2 L} \qquad \textbf{Equation 12-3}$$

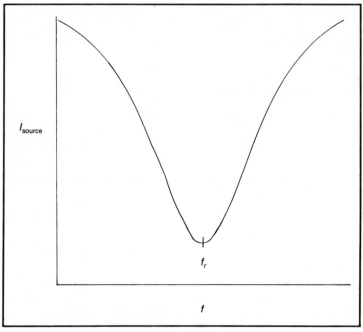

Fig. 12-4. Current versus frequency response of Fig. 12-3.

where C is the capacitance in farads;
             π is 3.14
             f is the frequency in hertz;
             L is the inductance in henrys.

In summary, the circuit impedance will be at its lowest at the resonant frequency in a series resonant circuit, while current will be
at its highest. The circuit impedance will be at its highest at the
resonant frequency in a parallel resonant circuit, while source current will be at its lowest.

    The intent of this chapter is to provide you with a well-rounded
background in electronics. In actuality, you will probably never need
to work with resonant circuits in industrial electronics. Nevertheless, resonant circuits are sometimes used for noise reduction if
the noise is definable relative to its frequency content. Generally
speaking, resonant circuits are more commonly used in radio-
frequency circuits for selecting or transmitting specific frequencies.
The tuner on your television or radio is an example of a tunable
resonant circuit.

Resonant circuits can also be used to select certain frequencies from sources containing random or mixed frequencies. Resonant circuits used for this purpose are usually called *filter circuits*.

INFORMATION CONCERNING
THE EQUATIONS USED IN THIS CHAPTER

You may find the equations used in this chapter difficult to understand if you do not have an extensive mathematical background. For the present, simply remember that this section exists if you ever need to work with resonance. If you wish to perform any resonance calculations, the following explanation of the mathematics involved should help.

Assume that the circuit shown in Fig. 12-1 has a capacitance value of 1 microfarad and an inductance value of 1 henry. If you wish to calculate the resonant frequency, the previous values will plug into the formula as follows:

$$f_r = \frac{1}{6.28 \sqrt{LC}} = \frac{1}{6.28 \sqrt{(1)(.000001)}} = \frac{1}{6.28 (.001)}$$

Note that the square root of the inductance and capacitance values must be found before multiplying by 6.28. Therefore, the final answer is:

$$\frac{1}{.0068} = 159.23 \text{ hertz}$$

In Equation 12-2, the frequency calculated from the previous example will be used. The Greek letter pi (π) is a constant used in many mathematical calculations. Its value is always 3.14. In this equation, pi and the frequency must be squared before they are multiplied by any of the other values. The values are calculated as follows:

$$L = \frac{1}{4\pi^2 f^2 C} = \frac{1}{(4)(9.8596)(25354.2)(.000001)}$$

This provides the following solution:

$$\frac{1}{.9999} \cong 1 \text{ henry}$$

Using the same values for Equation 12-3, the following is derived:

$$C = \frac{1}{4\pi^2 f^2 L} = \frac{1}{(4)\,(9.8596)\,(25354.2)\,(1)}$$

This provides the following solution:

$$\frac{1}{999970} \cong .000001 = 1 \times 10^{-6} = 1 \text{ microfarad}$$

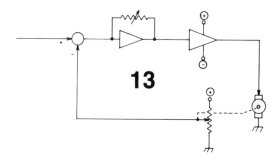

Diodes

EARLIER IN THIS BOOK, YOU LEARNED THE BASICS OF ELEC-
tron flow. You discovered that electrical current is actually a
flow of negative charge carriers called electrons. Electrons orbit
around the nucleus of an atom just as the earth orbits around the
sun. Electrons are held in their orbital paths by their attraction to
the positive nucleus, which contains the positive charge carriers
called protons. Electrons are attracted to the positive nucleus due
to a basic law of physics, which states unlike charges attract and
like charges repel. This same principle can be demonstrated by observing the attraction between two permanent magnets. The two
north poles of the magnets will repel each other, while a north and
south pole will attract. In the same way, the unlike charges of the
negative electron and the positive proton attract one another.

INTRODUCTION TO SOLID-STATE DEVICES

Because one electron (negative charge carrier) will equalize the
effect of one proton (positive charge carrier), a normal atom will
be balanced in reference to the number of electrons and protons.
For example, if an atom has eleven electrons, it will also contain
eleven protons. The end result is that the negative charge of the
electrons will be canceled out by the positive charge of the protons, and the atom will not present any external charge.

The ortibal paths of the electrons around the nucleus follow a definite pattern of circular shells, or rings. The maximum number of electrons in each shell is defined by the formula:

$$2 \ (n^2) \qquad \textbf{Equation 13-1}$$

In Equation 13-1, n is the shell number. For example, when considering the first shell, $n = 1$. Therefore:

$$2 \times (1 \times 1) = 2$$

The maximum number of electrons in the first shell of any atom is 2, regardless of the total number of electrons in that particular atom. (The exception is hydrogen; it has only one electron, so there is only one electron in the first shell.) Likewise, the total number of electrons possible in the second shell can be calculated with the previous formula:

$$2 \times (2 \times 2) = 8$$

The maximum number of electrons in the second shell of any atom is 8. Likewise, the maximum numbers of electrons in the third, fourth, and fifth shells are 18, 32, and 50 respectively.

In electronics, only the last shell of an atom is important, because all electron flow occurs with electrons from the last shell. The last shell of an atom is called the *valence shell*. If the valence shell contains the maximum number of electrons (according to Equation 13-1), the substance is said to be an insulator, because the electrons are rigidly bonded together. If the valence shell is far from being full, the electrons are easily loosened from their orbital bonds, and the substance is said to be a conductor.

The atoms in some types of crystalline substances (such as silicon) fill their valence shells by overlapping the orbital paths of neighboring atoms. In this manner, a single electron is actually shared by two atoms. For example, an isolated silicon atom contains four electrons in its valence shell. When silicon atoms combine to form a solid crystal, each atom positions itself between four other silicon atoms in such a way that the valence shells overlap from one atom to another. This causes each individual valence electron to be shared by two atoms, as shown in Fig. 13-1. By sharing the electrons from four other atoms, each individual silicon atom appears to have eight electrons in its valence shell. This effect of sharing valence electrons is called *covalent bonding*.

In its pure state, silicon is an insulator, because the covalent bonding rigidly holds all of the electrons leaving no free (easily loosened) electrons to conduct current. But consider the effect of injecting an impurity, with an atomic structure containing five electrons in its valence shell, into the pure silicon. The additional electron in the valence shell of the impurity cannot fit into the covalent-bonding pattern requiring four electrons in the valence shell. The result is a free electron that can readily move and conduct current.

Likewise, if an impurity is injected containing three electrons in its valence shell, the absence of the fourth electron needed for proper covalent bonding causes a free positive charge. (A free positive charge is another way of describing a hole, or absence of an electron.)

In both cases, a *semiconductor* has been created. The term semiconductor simply indicates that the substance is no longer a good insulator or a good conductor; it is somewhere in between. (The term semi-insulator would be just as accurate as the term semi-

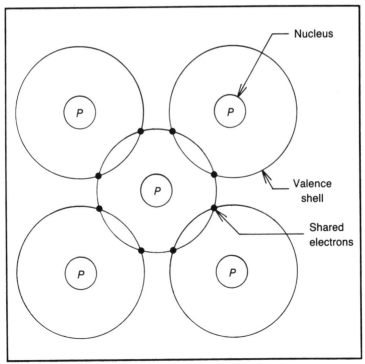

Fig. 13-1. Basic atomic structure illustrating covalent bonding.

conductor.) The process of injecting an impurity into a substance to create a semiconductor is called *doping*.

Besides the fact that the impure silicon becomes a semiconductor, it also possesses a unique property, depending upon whether it has been doped with a *pentavalent* impurity (an impurity with five electrons in its valence shell) or a *trivalent* impurity (an impurity with three electrons in its valence shell). Silicon doped with a pentavalent impurity will become *N*-type material. In opposition, doping silicon with a trivalent impurity creates a *P*-type material. *N*-type semiconductor material contains an excess of negative-charge carriers, and *P*-type material contains an excess of positive-charge carriers.

DIODE PRINCIPLES

When an *N*-type semiconductor is sandwiched with a *P*-type semiconductor, the resulting component is called a *diode*. A diode is a two-layer device that has an extremely low resistance to current flow in one direction and an extremely high resistance to current flow in the other. Because it is a two-layer device, it can also be considered a single-junction device, because there is only one junction between the *P* and *N* material, as shown in Fig. 13-2. A diode is often called a rectifier.

Ideally, you can consider a diode capable of passing current in only one direction. If the *P* side is positive relative to the *N* side by an amount greater than its forward threshold voltage (about .7 volt if silicon and .3 volt if germanium), the diode will freely pass current almost like a closed switch. If the *P* side is negative relative to the *N* side, virtually no current will be allowed to flow unless the device's breakdown voltage is achieved. If the reverse breakdown voltage is exceeded in a normal diode, the diode will be destroyed.

The *P* side of a diode is called the *anode*. The *N* side is called the *cathode*. (See Fig. 13-3.)

The principle behind diode operation is shown in Fig. 13-4. The diagram of a forward-biased diode demonstrates the operation of a diode in the forward-conduction mode (freely passing current). With the polarity, or bias, of voltages shown, the forward-biased diode will conduct current as if it were a closed switch.

As stated previously, like charges repel each other. If a positive voltage is applied to the *P* material, the free positive-charge carriers will be repelled and move away from the positive potential toward the junction. Likewise, the negative potential applied

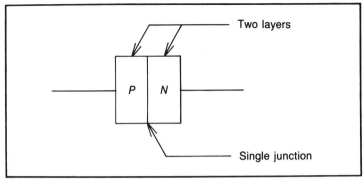

Fig. 13-2. Basic diode construction.

to the *N* material will cause the free negative-charge carriers to move away from the negative potential toward the junction.

When the positive and negative charge carriers arrive at the junction, they will attract (unlike charges attract) and combine. As each negative and positive charge carrier combine at the junction, a new negative and positive charge carrier will be introduced to the semiconductor material from the source providing the bias. As these new charge carriers enter the semiconductor material, they will move toward the junction and combine. Thus, current flow is established and will continue as long as the bias voltage remains.

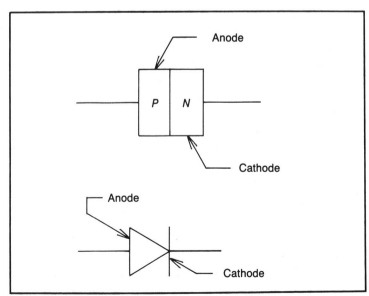

Fig. 13-3. Diode components and electrical symbol.

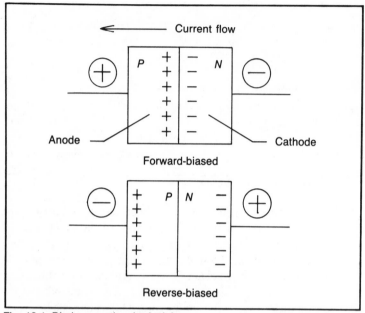

Fig. 13-4. Diode operational principles.

The forward threshold voltage must be exceeded before a forward-biased diode will conduct. The forward-threshold voltage must be high enough to loosen the charge carriers from their atomic orbit and push them through the junction barrier. With silicon diodes, this forward-threshold voltage is approximately .7 volt. With germanium diodes, the forward-threshold voltage is approximately .3 to .4 volt.

A forward-biased diode is limited to the amount of current it can withstand. This limit is based on the physical size and construction of the diode. Diode manufacturers specify the maximum forward current for a particular diode.

Figure 13-4 also illustrates a diagram of a reverse-biased diode. As may be expected, the opposite effect occurs if the *P* material is negative-biased relative to the *N* material. In this case, the negative potential applied to the *P* material attracts the positive charge carriers, drawing them away from the junction. Likewise, the positive potential applied to the *N* material draws the negative charge carriers toward it and away from the junction. This leaves the junction area depleted; virtually no charge carriers exist. Therefore, the junction area becomes an insulator, and current flow is inhibited.

The reverse-bias potential may be increased to the reverse-breakdown voltage for which the particular diode is rated. As in the case of the maximum forward-current rating, the reverse-breakdown voltage is specified by the diode manufacturer. The reverse-breakdown voltage is usually much higher than the forward-threshold voltage. A typical general-purpose diode may be specified as having a forward-threshold voltage of .7 volt and a reverse-breakdown voltage of 400 volts. Exceeding the reverse-breakdown voltage is destructive to a general-purpose diode. (Some manufacturers refer to the reverse-breakdown voltage as the peak-inverse voltage, or PIV.)

Diodes are commonly used to convert alternating current (ac) to direct current (dc). When four diodes are connected together and are used to redirect both the positive and negative alterations of ac to dc, this four-diode connection is called a *diode bridge*, or *bridge rectifier*. A single diode used for rectification is called a *half-wave rectifier*. These configurations can be demonstrated in a few common types of power-supply circuits.

Figure 13-5 shows a simple half-wave rectifier circuit. Common household voltage (120 Vac) is applied to the primary of a step-down transformer (T_1). The secondary of T_1 steps down the 120 V rms to 12 V rms. (You should calculate the turns ratio of T_1 to be 10:1. Transformer action is covered in Chapter 10.) The diode (D_1) will only allow the current to flow in the direction shown (from cathode to anode). Diode D_1 will be forward-biased during each positive half-cycle (relative to common). When the circuit current tries to flow in the opposite direction, the voltage bias across the diode will be reversed, causing the diode to act like an open switch. As shown in Fig. 13-6, this causes a pulsating voltage to be applied across the load resistor (R_L). Since common household ac voltage cycles at a 60-hertz frequency, the pulses seen across R_L will also be at 60 hertz. Figure 13-6 also shows the voltage waveform across D_1. During the positive half-cycle, the diode will drop the .7-volt forward-threshold voltage. The entire negative half-cycle will be dropped across D_1 while it is reverse biased.

Consider the amplitude of the voltage developed across R_L. As shown in Fig. 13-5, the secondary of T_1 is 12 Vac. The ac voltages are always assumed to be rms values unless otherwise stated. Therefore, the peak voltage output from the T_1 secondary is:

PEAK = (rms value) × 1.414 = (12) × 1.414 = 16.968 volts peak

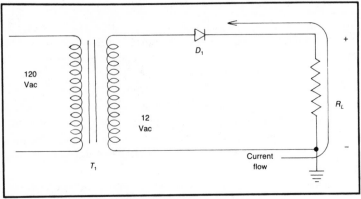

Fig. 13-5. Half-wave rectifier circuit.

This value can be rounded off to approximately 17 volts. Therefore, the T_1 secondary will output a 17-volt positive peak and a 17-volt negative peak to the circuit shown in Fig. 13-5. (You may want to review Chapter 9, *Alternating Voltage and Current*.) The negative half-cycles are blocked by D_1, allowing R_L to admit only the positive half-cycles. The actual peak voltage across R_L will be the 17-volt positive peak being supplied by the T_1 secondary minus the .7-volt forward-threshold voltage being dropped by D_1. In other words, 16.3-volt positive peaks will be applied to the load resistor R_L, as shown in Fig. 13-6.

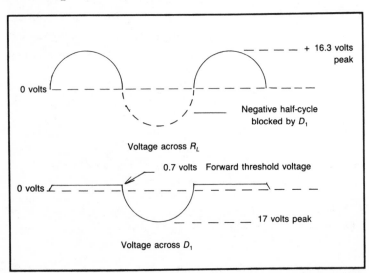

Fig. 13-6. Waveshapes across D_1 and R_L from circuit illustrated in Fig. 13-5.

A slight variation to the circuit in Fig. 13-5 is shown in Fig. 13-7. The addition of capacitor C_1 will make the circuit more practical for most applications. When the primary power is first applied to transformer T_1, the first positive half-cycle output from the secondary of T_1 will charge C_1 to the peak value seen across R_L. As shown in Fig. 13-7, C_1 charges to 16.3 volts at the peak of the positive half-cycle. Because C_1 and R_L are in parallel, the voltage across R_L will be the same as that across C_1.

The time required for C_1 to charge to the maximum (peak) level is determined by the circuit time constant, which is the series resistance multiplied by the capacitance value. In this circuit, the only thing in series with C_1 is the diode D_1. As soon as the forward-threshold voltage of D_1 is exceeded, D_1 resembles a closed switch. Therefore, the only appreciable resistance in series with C_1 above the forward-threshold voltage is the output impedance of the secondary winding in T_1. This impedance is normally considered negligible. Therefore, the resistance in series with C_1 is virtually zero. Any capacitance value multiplied by zero will give an answer of zero. Therefore, you can assume that the capacitor charges to the peak value instantaneously.

The time required for C_1 to discharge is quite different. Once C_1 has been charged in the polarity shown in Fig. 13-7, it cannot discharge through D_1 and the T_1 secondary, because D_1 will be reverse-biased. Therefore, it must discharge through the load resistor R_L. The discharge time constant is the total series resistance in the discharge path multiplied by the capacitance value.

The total series resistance in this case is R_L. Depending on the value of C_1 and R_L, the discharge time could be considerable.

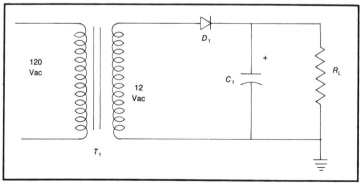

Fig. 13-7. Filter capacitor added to circuit illustrated in Fig. 13-5.

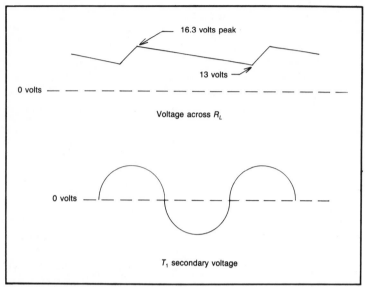

Fig. 13-8. Waveshape across R_L relative to the T_1 secondary voltage from the circuit illustrated in Fig. 13-7.

Therefore, the circuit illustrated in Fig. 13-7 can maintain, or filter, the pulsating dc across R_L and convert it into a relatively smooth, continuous dc voltage. Unfortunately, C_1 will discharge by some amount (during the negative half-cycle periods from the T_1 secondary), causing the voltage across R_L to decrease by some amount.

As shown in Fig. 13-8, the voltage across R_L does decrease to 13 volts before the next positive cycle has a chance to recharge C_1. This variation in the dc voltage level is called *ripple*. Ripple is undesirable in most cases. The amplitude of the ripple can be reduced in this circuit by increasing the discharge time constant (either by increasing the capacitance value of C_1 or by increasing the resistance value of R_L). You can consider ripple as being an ac component riding on the dc level. In other words, there is some percentage of periodic ac mixed with the dc voltage.

Capacitors used to filter out an ac component are commonly called *filter capacitors*. Capacitor C_1 in Fig. 13-7 is a filter capacitor.

Figure 13-9 shows another variation of the previous circuit. This circuit is identical to the previous one except for the addition of the inductor L_1. As you should recall from Chapter 10 on inductors, an inductor opposes a change in current and presents an op-

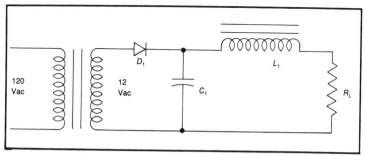

Fig. 13-9. Utilization of a choke (L_1) to reduce ripple.

position to ac called *inductive reactance.* Since L_1 is in series with R_L in this circuit, its inductive reactance to the ac component (ripple) is also in series with R_L. In contrast to the ac opposition, the dc opposition presented by L_1 is very small. Therefore, L_1 opposes the ac component but offers very little opposition to the dc component. If the inductance value of L_1 is properly chosen, most of the ripple will be dropped across L_1 while allowing most of the dc component to be dropped across R_L. Therefore, L_1 greatly reduces the percentage of ripple appearing across the load (R_L). Inductors used for this purpose are commonly called *chokes.*

Most modern electronic equipment requires a dc voltage source to operate properly. A dc voltage source is commonly called a *dc power supply.* The circuits shown in Figs. 13-7 and 13-9 are dc power supplies. Typically, the peak-to-peak ripple variations are less than 10 percent of the dc voltage. The load resistor (R_L) shown in the previous circuits is representative of the electronic circuit being powered by the power supply.

The power supplies shown in Figs. 13-7 and 13-9 are usable power supplies, but only half of the complete ac cycle is being utilized. The diode (D_1) used in these circuits is appropriately called a *half-wave rectifier.* However, it would be much more desirable to utilize the full ac cycle. The circuits shown in Figs. 13-10 and 13-11 are designed to accomplish this.

The full-wave bridge rectifier circuit shown in Fig. 13-10 is the most common type of ac rectifier. It consists of four diodes (D_1, D_2, D_3, and D_4).

Consider the operation of a bridge rectifier by referring to Fig. 13-12. When the transformer secondary outputs a half-cycle with the polarity shown, the current will follow the path shown by the arrows. (Remember, current always flows through a diode from cathode to anode.) With this polarity applied to the bridge, diodes

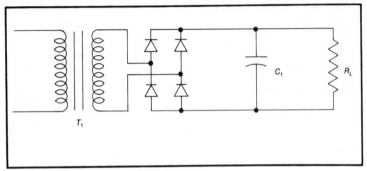

Fig. 13-10. Full-wave bridge rectifier.

D_1 and D_4 are forward-biased, while D_2 and D_3 are reverse-biased.

Figure 13-13 shows the current path when the polarity on the transformer secondary reverses. With this polarity applied to the bridge, diodes D_2 and D_3 are forward-biased, while D_1 and D_4 are reverse-biased. In either case, the current always flows through the load resistor (R_L) in the same direction, and the voltage polarity across R_L does not change. The switching action of the diode bridge actually turns the negative half-cycle upside-down.

Referring back to Fig. 13-7 and Fig. 13-9, note that C_1 is being recharged at a rate of 60 times per second, because the secondary output of T_1 is at 60 hertz. When a diode bridge is utilized to turn the negative half-cycle upside-down, it actually converts the 60 hertz ac to 120 hertz pulsating dc.

Referring to Fig. 13-14, note that the negative half-cycles from the T_1 secondary are turned upside-down by the diode bridge, causing the frequency to double. Due to the frequency-doubling ef-

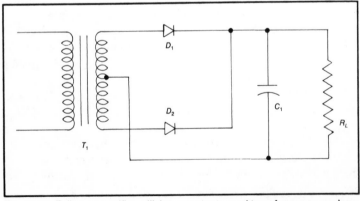

Fig. 13-11. Full-wave rectifier utilizing a center-tapped transformer secondary.

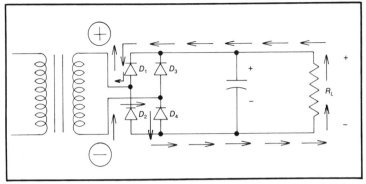

Fig. 13-12. Current flow through a full-wave bridge rectifier with indicated ac polarity.

fect, C_1 is recharged twice as often, or at a 120-hertz rate. This is the advantage of the circuit shown in Fig. 13-10. By recharging C_1 twice as often, it only has half the time to discharge through the load resistor (R_L). Therefore, the ripple is substantially reduced.

NOTE: You may be confused about the difference between pulsating dc and ac. As stated in previous chapters, ac is a voltage polarity and current reversal. With pulsating dc, there is a large ac component (ripple), but the current flow never changes direction through the load. In Fig. 13-14, note that the pulsating dc never crosses the zero reference line into the negative region.

Another type of full-wave rectification system is shown in Fig. 13-11. This circuit operates just as well as the full-wave bridge rec-

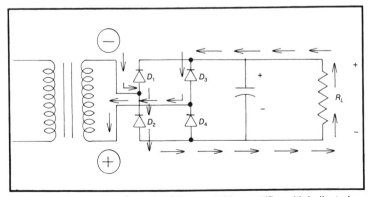

Fig. 13-13. Current flow through a full-wave bridge rectifier with indicated ac polarity.

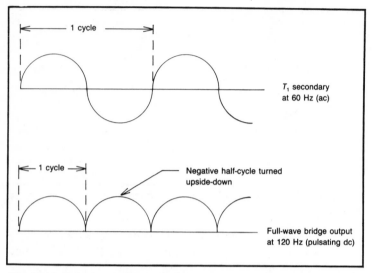

Fig. 13-14. Output waveshape of a full-wave rectifier relative to the T_1 secondary voltage from the circuit illustrated in Fig. 13-11.

tifier discussed previously, but it requires a center-tapped transformer. The secondary center tap becomes the power supply's common (reference potential).

As shown in Fig. 13-15, the two outputs from each side of the secondary are exactly opposite to each other in reference to the center-tap. (In other words, the two outputs are 180 degrees out of phase.) If one output is in the positive half-cycle, the other output must be in the negative half-cycle and vice versa. Thus, at virtually any point in time, one of the two outputs will be in the positive half-cycle.

The circuit in Fig. 13-11 is actually two half-wave rectifiers connected to the same load and filter capacitor. Because the two outputs from the T_1 secondary are opposite each other, if D_1 is forward-biased, D_2 will be reverse-biased. When D_1 becomes reverse-biased, D_2 becomes forward-biased. In actuality, two half-wave rectifiers are connected to form a full-wave rectifier. The end effect is the same as that achieved with the circuit in Fig. 13-10. However, to achieve the same dc voltage amplitude to the load, the transformer's rated secondary voltage in Fig. 13-11 must be twice the value of the transformer shown in Fig. 13-10. (This is because a 12-volt center-tapped secondary is actually only 6 volts from the center tap to either side of the winding.)

140

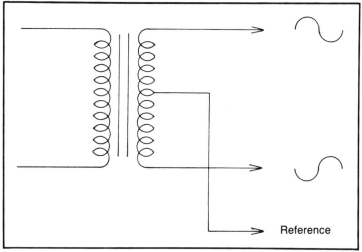

Fig. 13-15. Two secondary leads of a center-tapped transformer, 180 degrees out of phase.

ZENER DIODES

In some electronic circuits, it is critical to maintain a very precise power-supply voltage. The power supplies examined thus far are called *unregulated power supplies*. The output from an unregulated power supply can vary by significant amounts if its load changes (the load is the circuit or device to which the power supply is supplying power) or if the 120-Vac line voltage varies (common household power can vary by more than 10 percent). Thus, if a precise output voltage must be maintained from a power supply, the power supply must be regulated.

Zener diodes are used primarily as voltage regulator devices. A zener diode is a special type of diode manufactured to be operated in the reverse breakdown region. Every zener diode will be manufactured for a specific reverse breakdown voltage called the zener voltage. (The zener voltage will be specified by the manufacturer.)

Figure 13-16 shows a zener diode-regulated power supply. This zener diode (D_Z) is a 15-volt zener. If it is reverse-biased at a voltage higher than 15 volts, it will break down and maintain 15 volts across itself. The remaining voltage will be dropped across the series resistor R_S.

For example, assume that the unregulated dc voltage across C_1 in Fig. 13-16 is 18 volts. (Note that D_Z is reverse-biased in this

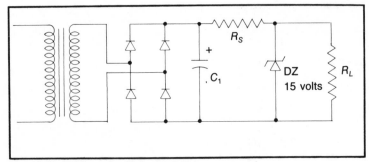

Fig. 13-16. Basic zener diode regulated power supply.

circuit.) Zener diode D_Z will break down at 15 volts, and the remaining 3 volts will be dropped across R_S. (Remember, the sum of the individual voltage drops must equal the applied voltage.) If the unregulated dc voltage across C_1 drops to 16 volts, the voltage across D_Z remains at 15 volts, but the voltage across R_S drops to only 1 volt. If the voltage across C_1 increases to 20 volts, the voltage across D_Z remains at 15 volts, while the voltage across R_S increases to 5 volts. Since the load resistance (R_L) is in parallel with the zener, the zener maintains a regulated 15 volts across the load regardless of the unregulated dc voltage fluctuations. (In reality, there would be a small voltage variation across the zener from one extreme of the raw dc voltage [the unregulated dc voltage in a power supply] to the other. In most applications, this variation is considered negligible.)

TESTING DIODES

All general-purpose diodes may be tested with an ohmmeter.

Modern electronic power supplies provide extremely good voltage/current regulation under varying loads and various ambient conditions (courtesy of Power-One, Inc.).

142

When the red (positive) side of the ohmmeter is connected to the anode and the black (negative) side is connected to the cathode, the diode under test is forward-biased, and the ohmmeter shows a low-resistance reading. When the leads are reversed, the diode is reverse-biased, and the ohmmeter indicates a very high, or infinite resistance. The ohmmeter should be switched to one of the lower resistance ranges for testing diodes.

The type of ohmmeter being used may affect this method of testing diodes. Newer types of DVMs have special range positions for testing diodes. They will not test diodes properly in any range position except the one intended for this purpose. If uncertain, you should refer to the user's manual for the particular ohmmeter being used.

A zener diode may usually be tested in the same manner as is a general purpose diode. If it tests the same as does a general purpose diode, it is probably good. The best test of a zener diode is to apply power to the circuit it is used in and measure the voltage across it. If this voltage reading is the correct zener voltage (as rated by the manufacturer), the zener is good. If the voltage

Bridge Leads				Results
(+)	(−)	(ac)	(ac)	
Black		Red		Low resistance
Black			Red	Low resistance
Red		Black		Infinite resistance
Red			Black	Infinite resistance
	Red	Black		Low resistance
	Red		Black	Low resistance
	Black	Red		Infinite resistance
	Black		Red	Infinite resistance

Fig. 13-17. Correct measurements of a diode bridge module.

across the zener is incorrect, disconnect the load it is powering and try again.

Referring to Fig. 13-16, an excessive load would cause an excessive current flow through R_S. In turn, the voltage drop across R_S would be excessive, causing the voltage drop across D_Z to drop below its zener voltage.

Many full-wave diode bridges will be constructed as single hybrid modules. These modules are usually rectangular-shaped, with four leads extending from the package. Two of the four leads will be marked with polarity signs (+ and −). These are the two dc output leads with the output polarity as indicated on the package. The other two leads are the ac input leads; thus, there is no polarity associated with them.

Diode bridge modules can also be tested with an ohmmeter. Figure 13-17 is a chart indicating the correct measurements of a diode bridge module. The four diode bridge leads are shown at the top of the chart. The terms BLACK and RED refer to the connection points of the black (negative) and red (positive) ohmmeter leads. The measurement results are shown at the right of the chart. (The correct low-resistance reading will depend upon the ohmmeter being used and the range chosen.)

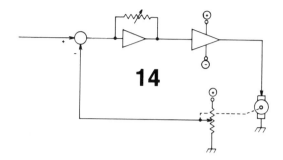

The Transistor

THE *TRANSISTOR* IS A SOLID-STATE, THREE-LAYER SEMICON-
ductor device. Figure 14-1 shows the basic construction of a
transistor and compares transistor construction to diode construction. Note that a diode contains only one junction, whereas a transistor contains two junctions. Because a transistor contains two
junctions, it is often referred to as a bipolar (dual-junction) device.
Figure 14-1 also shows that bipolar transistors can be constructed
in either of two configurations—NPN or PNP.

Bipolar transistors have three connection points, or leads, attached to them. These are called the *emitter*, the *base*, and the *collector*. The symbols for bipolar transistors are shown in Fig. 14-2.
The only difference between NPN and PNP symbols is the direction of the arrow in the emitter lead.

GAIN

Before you learn the principles behind the transistor operation,
you should understand the basic concept of *gain*. The most common types of gain you will encounter in the electronics field are
current gain, voltage gain, and *power gain.*

Gain is simply the ratio of the input (voltage, current, or power)
to the output (voltage, current, or power). For example, if you apply a 1-volt signal to the input of a circuit and that circuit increases
the amplitude of the 1-volt signal to 10 volts, there is a voltage gain
of 10.

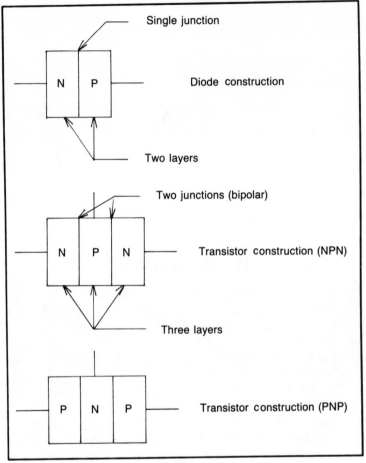

Fig. 14-1. Comparison between diode and transistor construction.

It is possible to have a gain of less than 1. For example, if the output of a circuit is only one-half of the input, the gain is .5. This means the output is being attenuated (reduced in amplitude) by a factor of 2.

TRANSISTOR PRINCIPLES

A *bipolar transistor* is an active device. In other words, it is capable of power gain. All of the devices previously discussed in this book (transformers, capacitors, resistors, etc.) are *passive* devices; they are not capable of power gain.

For example, a transformer is capable of providing voltage gain

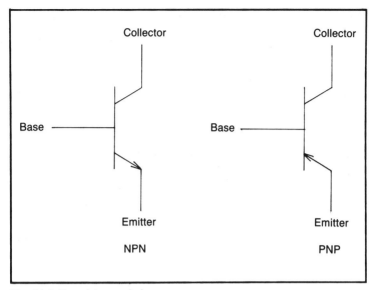

Fig. 14-2. Transistor symbols.

if it is a step-up transformer, but the secondary current is reduced by the turns ratio. Since power is equal to voltage times current, the primary-to-secondary power transfer in a transformer is always approximately 1 (actually, with the small transformer losses, the power transfer is slightly less than 1). Therefore, a transformer is considered a passive device.

In understanding transistor operation, you should always consider a transistor as being a current device. Regardless of the conversions that take place within a transistorized circuit, the transistor itself is only capable of providing current gain. Depending on the configuration of the transistor in a circuit, this current gain can be converted to a subsequent voltage and/or power gain.

The parameter (component specification) that defines the amount of current gain possible in a particular type of transistor is called *beta*. For example, if a certain type of transistor is specified by the manufacturer to have a beta of 100, this means the current flowing through the collector lead will be 100 times greater than the current flowing through the base lead (provided that the circuit in which the transistor is inserted has been properly designed). In many specification sheets, beta (current gain) is often referred to as hfe.

Equation 14-1 defines the relationship between base current and collector current:

$$I_c = I_b \text{ (beta)} \qquad \textbf{Equation 14-1}$$

To avoid confusion, consider the transistor illustrated in Fig. 14-3 as being an ideal (perfect) transistor. Note that the base-to-emitter junction (PN) is simply a diode junction. If a 10-volt positive potential is applied to the base lead with the emitter lead going to common (through resistor R_e), the diode junction will be forward-biased, and current will begin to flow from the emitter to the base. The amplitude of this current flow will be proportional to the input impedance at the base of the transistor circuit. For discussion purposes, assume the base current (I_b) to be 100

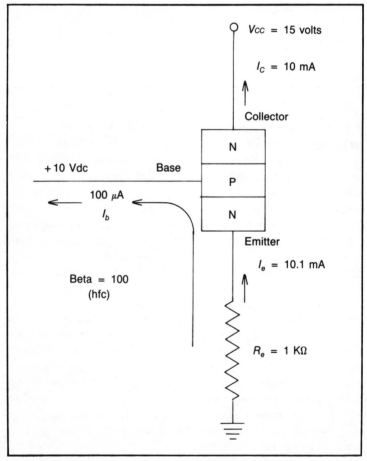

Fig. 14-3. Basic transistor operational principles.

microamps. If this transistor was specified to have a beta of 100, the current flow through the collector lead (I_c) would be:

$$I_c = I_b \text{ (Beta)} = (100 \text{ microamps}) (100) = 10 \text{ milliamps}$$

In other words, the collector current is 100 times greater than the base current.

In Fig. 14-3, note that both the collector current and the base current must flow through the emitter lead and resistor R_e. Therefore, the emitter current is simply the sum of the base current and the collector current. The base current will be so small compared to the collector current that it can be considered negligible. Therefore, for all practical purposes, the emitter current can be assumed to be the same as the collector current. This is stated in Equation 14-2:

$$I_e = I_c \qquad \qquad \textbf{Equation 14-2}$$

The circuit in Fig. 14-4 is the same as that shown in Fig. 14-3 except that the proper transistor symbol is shown. Using Ohm's law, the voltage dropped across resistor R_e can be calculated as follows:

$$E_{R_e} = I_e R_e = (10 \text{ mA}) (1 \text{ k}\Omega) = 10 \text{ volts}$$

NOTE: On schematic diagrams and in technical information, transistor voltages will be listed for troubleshooting purposes. These voltages are appropriately called the *emitter voltage*, the *collector voltage,* and the *base voltage.* Unless otherwise specified, emitter, base, and collector voltages will always be measured in reference to the circuit common.

The voltage of +10 volts dropped across R_e is called the *emitter voltage.* In reference to circuit common, this is the ideal voltage you would measure at the emitter lead. Note this is simply the voltage dropped across R_e.

The *collector voltage* (in reference to circuit common) is the applied source voltage (*VCC*).

The *base voltage* is the +10 volts mentioned previously.

The voltage drop between the collector and the emitter is in series with the voltage drop across R_e (See Fig. 14-4). Therefore, since the sum of the voltage drops must equal the applied source voltage (Vcc), the voltage drop between the collector and emitter must be:

$$E_{CE} = VCC - E_{R_e} = 15 - 10 = 5 \text{ volts}$$

Until now, you have examined transistor operation based on ideal conditions. Referring to Fig. 14-5, the base-to-emitter junction can be considered equivalent to a simple diode junction. As explained previously, the base-to-emitter junction is forward-biased when a voltage of +10 volts is applied to the base. Thus, there will be a forward threshold voltage drop across the base-to-emitter junction. With silicon transistors (most modern transistors are silicon), this forward threshold voltage will be approximately .7 volt. With germanium transistors, it will be approximately .3 volt.

In Fig. 14-6, the forward threshold voltage of the base-to-emitter junction has been taken into consideration. Notice that the

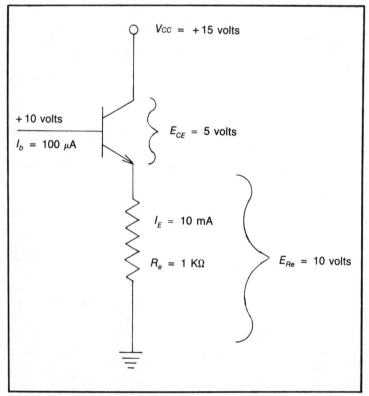

Fig. 14-4. Basic transistor circuit.

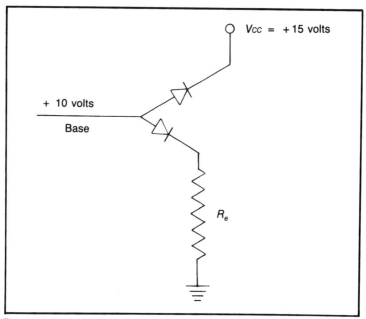

Fig. 14-5. Equivalent circuit of Fig. 14-4.

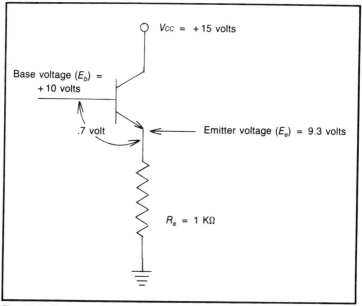

Fig. 14-6. Basic transistor circuit illustrating the forward voltage drop across the base to emitter junction.

sum of the emitter voltage and the forward threshold voltage of the base-to-emitter junction equals the applied +10 volt potential to the base. Equations 14-3 and 14-4 define the relationship between the base and emitter voltages:

$$E_e = E_b - .7 \text{ (silicon transistors)} \qquad \textbf{Equation 14-3}$$

$$E_e = E_b - .3 \text{ (germanium transistors)} \qquad \textbf{Equation 14-4}$$

Refer to Fig. 14-6. According to Ohm's law, the emitter current must be the emitter voltage (voltage across R_e) divided by the resistance value of R_e. This is stated in Equation 14-5:

$$I_e = \frac{E_e}{R_e} \qquad \textbf{Equation 14-5}$$

Therefore, the emitter current of the circuit illustrated in Fig. 14-6 is:

$$I_e = \frac{E_e}{R_e} = \frac{9.3 \text{ volts}}{1 \text{ k}\Omega} = 9.3 \text{ mA}$$

Since Equation 14-2 states the emitter current can be assumed to be the same as the collector current, the collector current in Fig. 14-6 is also 9.3 mA.

Using a derivation of Equation 14-1, the actual base current can be calculated by dividing the collector current by beta:

$$I_b = \frac{I_c}{\text{beta}} = \frac{9.3 \text{ mA}}{100} = 93 \ \mu\text{A}$$

Because you now know the current flow through the base lead (93 μA) and the voltage applied to the base (10 volts), you can use Ohm's law to calculate the input impedance at the base of the transistor:

$$Z_b = \frac{E_b}{I_b} = \frac{10 \text{ volts}}{93 \ \mu\text{A}} = 107 \text{ k}\Omega$$

In other words, the +10-volt signal applied to the base imparts a load (input impedance) of 107 kohms. Normally, for input-impedance calculations, the .7-volt forward threshold voltage between the base and emitter (in silicon transistors) is subtracted from the applied base voltage. This is done because the .7-volt drop is reasonably constant regardless of the amplitude of the applied base voltage. Once you perform this subtraction, the true input impedance calculation is:

$$Z_b = \frac{E_b - .7}{I_b} = \frac{9.3 \text{ volts}}{93 \ \mu A} = 100 \text{ k}\Omega$$

This value (100 kohm) is equal to the value of R_e (1 kohm) multiplied by beta (100). This much easier way to calculate the input impedance is shown in Equation 14-6:

$$Z_b = (\text{beta}) \ R_e \qquad \textbf{Equation 14-6}$$

where Z_b is the input impedance of the base;
R_e is the emitter resistance.

Also, notice that if you subtract the forward-threshold voltage from the applied base voltage, the voltage across the emitter resistor (R_e) is equal to this compensated base voltage. In other words, this circuit has a voltage gain of 1. Thus, if you divide the input (base) voltage by the output (emitter) voltage, it will always be approximately equal to 1. Equation 14-7 shows this relationship:

$$E_b = E_e \qquad \textbf{Equation 14-7}$$

where E_b is the base voltage;
E_e is the emitter voltage.

In essence, the circuit illustrated in Fig. 14-6 amplifies the 93 μA of base current to 9.3 mA of emitter current while maintaining the same voltage on the emitter as that on the base. This constitutes a power gain. Because power is equal to the voltage times the current, the power delivered to the base is:

$$P_b = I_b \, E_b = (93 \ \mu A) \ (9.3 \text{ volts}) = 865 \ \mu W$$

Likewise, the power delivered to the emitter is:

$$P_e = I_e \, E_e = (9.3 \text{ mA}) \ (9.3 \text{ volts}) = 86.5 \text{ mW}$$

Notice that the power delivered by the emitter is 100 times greater than the power delivered to the base. This is the same value as beta. Equation 14-7 shows this relationship:

$$A_p = \text{beta} \qquad \textbf{Equation 14-7}$$

where A_p is the power gain.

Thus far, you have examined transistor operation based on quiescent (inactive, but still operating) conditions. But in most cases, the applied voltage to the base of a transistor is a dynamic (moving) signal. The emitter voltage in Fig. 14-6 will follow (or duplicate) any voltage fluctuations that occur on the base as long as the design parameters of the circuit are not exceeded. An example of exceeding the circuit design parameters would be to allow the base voltage to exceed V_{CC} or fall below the .7-volt forward-threshold voltage.

IMPEDANCE MATCHING

To understand the practical value of the circuit shown in Fig. 14-6, refer to Fig. 14-7, which shows a signal source with a high-output impedance. The output impedance (R_s) is achieved through the placement of a resistor in series with the actual signal voltage.

If R_s is 9 kohms and the source is connected to a 1-kohm load (R_e), a simple series voltage divider is established. When this occurs, 90 percent of the signal voltage will be dropped across R_s, while only 10 percent will be dropped by R_e. Ideally, the entire signal voltage (10 volts peak-to-peak) should be dropped across R_e.

Figure 14-8 shows how this can be accomplished with the addition of a few modifications to the previous circuits. Notice the source is a 10 volt peak-to-peak ac signal. If you applied this signal directly to the base of a transistor circuit, such as the one shown in Fig. 14-6, you would drive the base negative during the negative half-cycles of the ac signal. This is not desirable if you wish to amplify the entire waveshape. Therefore, the transistor must be biased at some dc level with sufficient amplitude to ensure that the base-to-emitter junction is never reverse-biased by the signal voltage.

This is accomplished by R_1 in Fig. 14-8. If the value of R_1 is chosen to hold the base at approximately $+7$ volts, the negative 5-volt peaks of the signal will only drive the base voltage down to

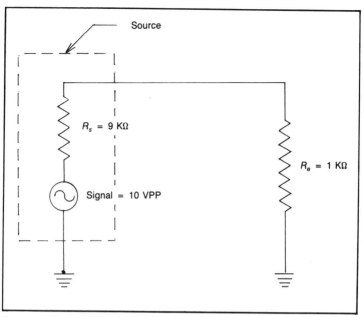

Fig. 14-7. Example of excessive loading to a signal source.

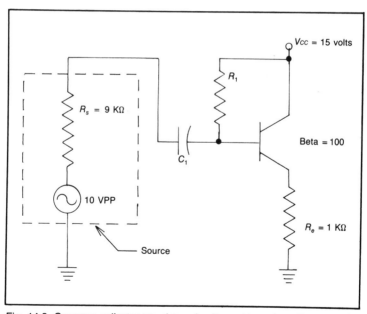

Fig. 14-8. Common collector transistor circuit used to reduce the excessive loading illustrated in Fig. 14-7.

155

approximately +2 volts (+7 volts −5 volts = +2 volts).

Likewise, during the positive 5-volt signal peaks, the base voltage will be driven up to approximately +12 volts and will not exceed V_{CC}.

Figure 14-9 shows the signal voltage, base voltage, and emitter voltage levels. Notice that the base voltage is riding on a dc offset of +7 volts, but the ac component still has a peak-to-peak amplitude of 10 volts. The emitter voltage is essentially the same as the base voltage with the exception of the .7-volt forward-threshold voltage. The value of C_1 is added to keep the dc base bias from being applied to the source. If the value of C_1 is properly chosen, it will easily pass the ac component but block the dc component. (A capacitor used for this purpose is generally called a *coupling capacitor.*)

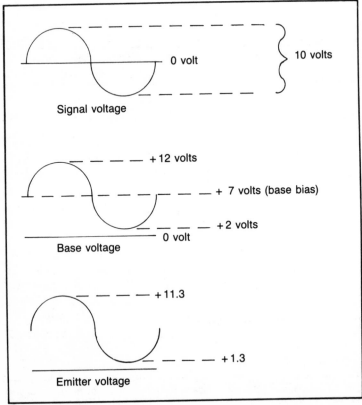

Fig. 14-9. Waveforms from the circuit shown in Fig. 14-8.

The important point of the circuit shown in Fig. 14-8 is that the signal source sees the input impedance at the base of the transistor instead of the 1-kohm resistance of R_e. This input impedance will be approximately equal to the value of R_e multiplied by beta (Equation 14-6). In this case, the input impedance will be:

$$Z_b = (1 \text{ kohm}) (100) = 100 \text{ kohms}$$

Therefore, only about 8 percent of the signal will be dropped across R_s, and the vast majority of the signal will appear across R_e. As you may recall, if the signal source had been applied directly to R_e, as shown in Fig. 14-7, 90 percent of the signal would have been dropped across R_s. Impedance matching is actually the process of providing enough current gain so that a high-impedance source can be applied to a low-impedance load without an appreciable loss.

NOTE: In Fig. 14-8, R_1 will affect the actual input impedance, causing it to be somewhat lower than Equation 14-6 would indicate. For precision, the resistance value of R_1 should be considered in parallel with the calculated base input impedance as shown in Equation 14-6.

THE COMMON COLLECTOR CONFIGURATION

The transistor circuit you have examined thus far is called a *common collector configuration*. In this configuration, the input is applied to the base of the transistor, and the output is obtained from the emitter.

To summarize, this is what you should have learned by now about the common collector configuration:

- The base voltage is considered equal to the emitter voltage.
- The input impedance at the base is equal to beta multiplied by the emitter resistance (in parallel with any base bias resistors).
- The collector current is equal to the emitter current.
- The collector current is equal to beta multiplied by the base current.
- The voltage gain is approximately equal to 1.
- The power gain is equal to beta.

The common collector configuration is used primarily for im-

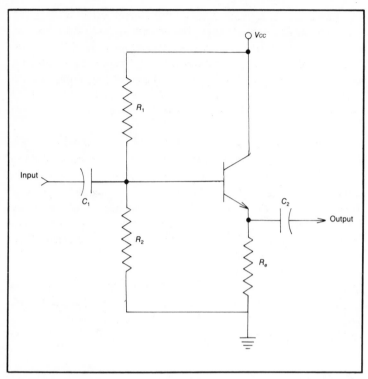

Fig. 14-10. Variation to the basic common collector circuit for providing improved stability.

pedance matching. A typical common collector stage is shown in Fig. 14-10. In this illustration, R_1 and R_2 form a stable voltage divider, which sets up the correct dc base bias to keep the base from becoming negative when ac signals are amplified. The C_1 keeps the dc base bias from being applied to the signal source. The C_2 keeps the dc offset on the output from being applied to the next stage (if applicable).

In Fig. 14-10, the actual input impedance would be equal to the value of R_1 in parallel with the value of R_2 in parallel with the product of R_e times beta.

THE COMMON EMITTER CONFIGURATION

Although the common collector configuration is very popular, the *common emitter configuration* is the most widely used transistor configuration. This is because the common emitter configuration is capable of voltage gain in addition to current gain.

Figure 14-11 shows a typical common emitter transistor circuit. Notice that the output is being taken from the collector, and a collector resistor (R_c) has been added. This circuit can be considered essentially the same as the typical common collector circuit. The input impedance is equal to beta multiplied by the emitter resistor (R_e) in parallel with any base bias resistors (R_1 and R_2). The emitter voltage is equal to the base voltage minus the forward-threshold voltage of the base-to-emitter junction. The collector current is equal to the product of the base current times beta. The emitter current can be considered equal to the collector current.

One major difference between the common collector and common emitter configuration concerns the voltage gain. As you may recall, the voltage gain of the common collector configuration is approximately equal to 1. With the common emitter configuration shown in Fig. 14-11, the voltage gain is equal to the ratio of the emitter resistor (R_e) to the collector resistor (R_c). Equation 14-8 defines this relationship:

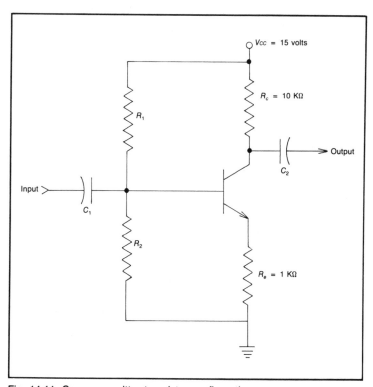

Fig. 14-11. Common emitter transistor configuration.

$$A_e = \frac{R_c}{R_e}$$

Equation 14-8

where A_e is the voltage gain.

Consider the reason why this relationship exists. Since the emitter current is considered essentially the same as the collector current, a variation of the collector current through the 10-kohm collector resistor (R_c) will cause a corresponding voltage drop 10 times higher than the same current variation will cause through the 1-kohm emitter resistor (R_e).

For example, using Ohm's law, a 1-milliamp current flow through a 10-kohm resistor will cause a 10-volt drop. A 1 milliamp current flow through a 1-kohm resistor will cause a 1-volt drop. Likewise, any current deviation through the 10-kohm resistor will cause a voltage drop ten times higher than the same current deviation would produce through the 1-kohm resistor.

As the emitter current is deviated by the base current deviations, the same current deviations are in the collector, causing a consequential voltage deviation across the collector resistor (R_c) 10 times the amplitude of that across the emitter resistor (R_e). Therefore, the actual voltage gain of this transistor stage is determined entirely by the ratio of the collector and emitter resistance values.

Referring to Fig. 14-11, assume that there is 1-milliamp of current flow through the emitter leg. This 1-milliamp current flow though the 1-kohm emitter resistor (R_e) will cause the emitter voltage to be 1 volt. The same 1 milliamp of current flow through the 10-kohm collector resistor (R_c) will cause 10 volts to be dropped across it.

It is important for you to realize at this point that the 10 volts dropped by the collector resistor (R_c) will not be the collector voltage. Since the collector is referenced to the circuit common, the collector voltage will be the sum of the emitter voltage and the voltage drop between the emitter and collector. In other words, the voltage drop across the collector resistor (R_c) is subtracted from the applied source voltage (VCC), and the difference is the voltage that will appear between the collector and circuit common. In this case, because the collector resistor (R_c) is dropping 10 volts, and $VCC = 15$ volts, the collector voltage will be the remaining 5 volts.

Referring to the previous example, assume that the emitter current is reduced to .5 milliamp. The emitter voltage will then be equal to .5 milliamp multiplied by 1 kohm, or .5 volt. The voltage drop across the collector resistor will be .5 milliamp multiplied by 10 kohms, or 5 volts. The collector voltage will be the applied collector voltage (VCC = +15 volts) minus the 5 volts dropped by the collector resistor, or 10 volts.

Notice that a .5-volt change in emitter voltage caused a 5-volt change in the collector voltage. The voltage deviation of the collector is 10 times greater than the voltage deviation of the emitter. Because the voltage deviation of the emitter is considered to be the same as the voltage deviation of the base, this circuit provides a voltage gain of 10.

Another interesting point is that, when the base voltage is reduced (to reduce the emitter voltage), the collector voltage increases. Consequently, if you increase the base voltage, the collector voltage will decrease. In other words, the collector output of a common emitter configuration is inversely proportional to the input. You can also say the output of a common emitter stage is 180 degrees out of phase with the input.

Figure 14-12 shows the relationship between a signal voltage input and output for the circuit shown in Fig. 14-11. The output signal is inverted and 10 times greater in amplitude than the input.

The power gain of a common emitter stage is the voltage gain (A_e) multiplied by the current gain (A_i). This is logical, because the basic equation for calculating power is $P = IE$.

In summary, all of the basic characteristics of the common collector configuration are applicable to the common emitter configuration with the exception of the following:

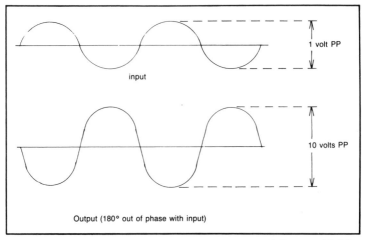

input

10 volts PP

1 volt PP

Output (180° out of phase with input)

Fig. 14-12. Output versus input waveshapes of the circuit illustrated in Fig. 14-11.

- The power gain is equal to the voltage gain multiplied by the current gain.
- The voltage gain is the ratio of the emitter resistance value to the collector resistance value.

THE EFFECT OF LOADING THE OUTPUT OF A TRANSISTOR STAGE

You have already seen the effect of trying to apply a high-impedance signal source to a low-impedance load. Most of the signal will be dropped across the internal source impedance. The primary function of the common collector configuration is to provide a high-input impedance to greatly reduce this undesirable loading effect.

In addition to the input impedance considerations, the load, as seen by the transistor stage output, can also affect the circuit operation. In Fig. 14-13, resistor R_L represents a load as seen by the

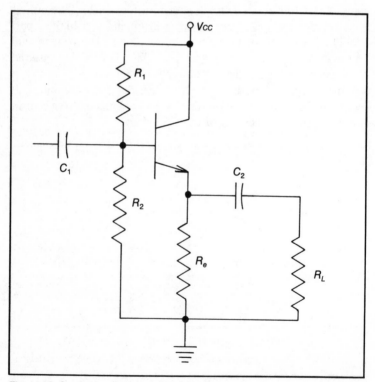

Fig. 14-13. Common collector configuration illustrating input and output coupling capacitors with an output load.

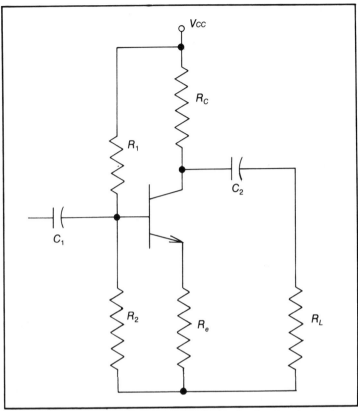

Fig. 14-14. Common emitter configuration illustrating input and output coupling capacitors with an output load.

output of a common collector circuit. Due to coupling capacitor C_2, the dc conditions of the circuit will not be affected because C_2 blocks any dc current from flowing through R_L.

On the other hand, if an ac signal is present at the output, the ac signal will be passed by C_2. This causes R_L to appear to be in parallel with R_e to the ac signal. Therefore, from an ac point of view, the emitter resistor (R_e) appears to be the equivalent parallel resistance of R_L and R_e. It is this equivalent parallel resistance that must be used in Equation 14-6 for the calculation of the ac input impedance.

As a general rule-of-thumb, if the output load impedance (R_L) is greater than 10 times the value of the emitter resistor, it can usually be considered negligible.

In Fig. 14-14, an output load (R_L) has been inserted on a common emitter stage. This load does not affect the input impedance of the stage (as was the case in the previous common collector circuit), but it will affect the ac voltage gain (and consequent power gain) of the circuit. To an ac output signal, resistors R_L and the collector resistor (R_c) appear to be in parallel. This equivalent parallel resistance is what the ac signal sees as the collector resistance. Therefore, since the ac signal gain is determined by the ratio of the emitter resistor (R_e) and the collector resistance, an output load can substantially reduce the voltage gain (if its impedance is low enough). Once again, if the output load impedance (R_L) is greater than 10 times the value of the collector resistor (R_c), it can usually be considered negligible.

OTHER TRANSISTOR CIRCUIT CONFIGURATIONS

There are many other types of transistor circuit configurations. Although these types are too numerous to be included in this book, you should be able to analyze the basic circuit operations with the knowledge you have gained by studying the two most common configurations (common emitter and common collector). Keep in mind that all transistors operate in the same basic manner, regardless of the circuit configuration. Also, the testing and troubleshooting procedures described later in this chapter are applicable to all transistor configurations.

NPN AND PNP TRANSISTORS

Basically, you may think of NPN and PNP transistors as being identical except that PNP transistors are normally used in conjunction with negative power supplies. So far in this chapter, you have examined circuits that use NPN transistors. If the applied source voltage (Vcc) happened to be negative rather than positive, a PNP transistor would have to be used.

Often, NPN and PNP transistors are used in complementary circuits in which the positive half-cycle of an ac waveshape is amplified by the NPN transistor and the negative half-cycle is amplified by the PNP transistor. In this type of circuit, the transistors must be matched according to operational characteristics so the output is not distorted (changed from its original form).

TRANSISTOR TYPES

Numerous types of transistors are available. Many different physical shapes and sizes are needed to accommodate variations

in power dissipation, gain, frequency, and maximum voltage requirements. If you will be working with transistors (or any other
type of solid-state device), you should obtain several manufacturer's
specifications and cross-reference manuals. These manuals will provide you with necessary information for transistor replacement purposes, along with definitions and product information. Your local
electronics sales and servicing stores can provide you with these
manuals or information on how to obtain them.

TROUBLESHOOTING TRANSISTORIZED CIRCUITS

Statistically speaking, semiconductor devices are much more
susceptible to failure than are most other components. (Semiconductor devices include diodes, transistors, integrated circuits, thyristor devices, etc.) There are several reasons for this. For one,
semiconductor devices can fail very rapidly if their maximum parameter ratings are exceeded. For example, if you exceeded the
maximum current rating for a particular diode, it can be destroyed
in just a few milliseconds. In contrast, electronic devices such as
resistors or transformers can usually handle a transient overload
without failing.

Another reason for more frequent failures of semiconductor
devices is a device parameter known as the *thermal cyclic curve*.
The thermal cyclic curve defines the number of times a semiconductor device can reach a certain operating temperature (in terms
of power dissipation) before it is likely to fail. Therefore, if a semiconductor is cycled through radical temperature changes for a considerable period of time, it will probably fail.

The remainder of this chapter will concentrate on testing and
troubleshooting bipolar transistor circuits. It is assumed at this point
that you already know how to test the associated passive devices
such as resistors, capacitors, transformers, and diodes.

TESTING TRANSISTORS OUT-OF-CIRCUIT

In some cases, it may be easy to remove a transistor from its
associated circuit for testing purposes. This is normally the case
if the transistor is mounted in a plug-in socket. You may also wish
to test a new transistor before using it as a replacement for a defective one. There are several good methods of testing transistors out-
of-circuit.

The best out-of-circuit transistor testing method involves the
use of a specialized transistor tester. Most transistor testers are

capable of testing transistors in or out of circuit. When performing out-of-circuit testing, these units can also provide valuable transistor parameter data for matching complementary transistor pairs or designing of specialized transistor circuits.

Transistor testers are manufactured by many different test-equipment manufacturers at various price ranges, depending on available options. Detailed operational manuals, with easily understood instructions regarding various testing methods, are provided with these units.

Although the use of specially designed transistor testers is the preferred out-of-circuit transistor testing method, the most common out-of-circuit testing method is performed with an ohmmeter. Refer back to Fig. 14-1. Note that a transistor consists of two diode junctions. These junctions can be tested with an ohmmeter in the same manner as any common diode.

For example, in Fig. 14-3, if the NPN transistor were removed from the circuit, the red (positive) lead of the ohmmeter could be placed on the base of the transistor, and typical low-resistance diode junction readings could be obtained on the ohmmeter by touching the black (negative) lead to either the collector or emitter leads. If you connected the black (negative) lead to the base and touched the red (positive) lead to either the emitter or collector, the junctions would be reverse-biased, resulting in a reading of infinite resistance. Obviously, if the transistor in question is a PNP, the ohmmeter readings obtained in the previous tests will be exactly opposite.

In addition to testing the two diode junctions, one final test should be performed with the ohmmeter. The emitter-to-collector resistance should be checked and should read virtually infinite in either orientation of the ohmmeter leads (this holds true for either NPN or PNP transistors).

Testing a transistor with an ohmmeter is a reasonably good test, but in rare cases, a transistor may fail when placed in an actual circuit and exposed to higher voltage or current conditions. The only way to test for this type of failure is to test the transistor under actual in-circuit operation.

TESTING TRANSISTORS IN-CIRCUIT

A variety of in-circuit transistor testers are available from test-equipment manufacturers. These units normally apply a small bias together with a signal voltage while monitoring the transistor in question for voltage or current gain. If gain is detected, the transistor is assumed good.

Transistor testers of this type are not 100 percent reliable because many transistorized circuit designs will interfere with the proper operation of these units. Nevertheless, they provide very rapid transistor testing and are adequate for most circuits.

The remainder of this section concentrates on in-circuit transistor testing and troubleshooting using a common VOM or DVM. The following methods and procedures are highly reliable and easily learned with a little practice. Although NPN transistors are used for discussion purposes, the same principles apply to PNP transistors. The only difference with PNP transistors concerns the polarity reversal.

For in-circuit transistor testing with a VOM or DVM, you need to understand basic transistor operation described earlier in this chapter. The most difficult part of in-circuit testing is determining if the circuit fault is due to the transistor or to some external fault which may cause the transistor to appear bad.

Basically, there are three possible faults that can cause a transistorized circuit to operate incorrectly. A fault can originate from the input source (preceding circuit or device), causing a loss of signal or a dramatic change in the normal input bias voltage. A fault can exist within the transistorized circuit (the transistor itself or any associated components). A fault can exist in the output load (succeeding circuit or device). By analyzing the transistor voltages from the circuit in question, you should be able to determine the fault area.

Many schematic diagrams show the correct transistor voltages in a properly operating circuit. If a schematic diagram of this nature is available, troubleshooting becomes much easier.

For example, assume you measured the transistor voltages in a defective circuit and compared them to the correct voltages appearing on the schematic diagram. If the base voltage was much lower than normal, then you would also expect the emitter voltage to be much lower and the collector voltage to be higher (except in a common collector configuration in which the collector voltage is always V_{CC}). These erroneous voltage readings do not necessarily indicate the transistor is defective. In this hypothetical case, the low base voltage indicates something defective in the base bias circuit or the preceding circuit. Therefore, you should direct your attention to these areas.

Consider another hypothetical troubleshooting example. Assume you measured the transistor voltages and discovered the base voltage was correct but the emitter voltage was considerably greater than the base voltage. This indicates the transistor is probably defective, because the emitter voltage should be slightly less than the base voltage due to the forward-threshold voltage drop of the base-to-emitter junction.

Consider one final troubleshooting example. The circuit in question is a common emitter configuration, as shown in Fig. 14-14. After measuring the transistor voltages, you discover the collector voltage is much higher than the correct voltage indicated on the schematic diagram. It also happens to be the same as the V_{CC} voltage. The base and emitter voltages are high enough to indicate that some value of collector current should be flowing. But since the collector voltage is the same as the V_{CC} voltage, there cannot be a voltage drop across the collector resistor (R_c). This indicates a complete loss of current flow through the collector or a shorted collector resistor. Most likely, the transistor collector is open. At this point, you should remove the transistor from the circuit and perform the out-of-circuit tests to verify that it is defective.

Unfortunately, you will probably not be able to obtain good schematic diagrams and voltage listings for every transistorized circuit you have to troubleshoot. The following methods are effective and can be used without any specific information about the circuit in question.

Before applying power to the circuit in question, some preliminary checks can be made with an ohmmeter to help speed up the troubleshooting process. After setting the ohmmeter to its lowest resistance range, check for any shorts (zero resistance) between the emitter-to-base, emitter-to-collector, and base-to-collector junctions. If any extremely low resistance readings are obtained, the cause of the low resistance should be isolated before applying any circuit power. If the reason for the low resistance reading is not readily apparent, you may have to remove the transistor from the circuit to help isolate the fault to either the transistor or the associated circuitry.

If everything looks reasonable with the in-circuit ohmmeter checks, power may be applied to the circuit. Unless you are already familiar with the particular transistor in question, you will probably have to refer to a manufacturer's cross reference manual to obtain the lead configuration and transistor type (either NPN or PNP).

Referring to Fig. 14-15, assume you measured these voltages on the base, the emitter, and the collector of the transistor in ques-

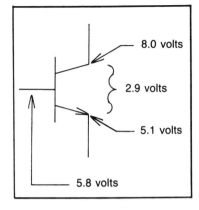

8.0 volts

2.9 volts

5.1 volts

5.8 volts

Fig. 14-15. Believable in-circuit transistor voltages.

tion. Without knowing anything about the associated circuit, you can make several determinations about the transistor's operation. The base-to-emitter junction appears to be good, because the emitter voltage is essentially the same as the base voltage minus the 0.7-volt forward-threshold drop. The emitter-to-collector junction is not shorted or breaking down, because there is a 2.9-volt potential difference between the emitter and collector.

Turn off the applied power to the circuit, and disconnect the base lead of the transistor from the associated circuit. Reapply the circuit power. Referring to Fig. 14-16, with the base lead open, the full V_{CC} supply voltage should appear at the collector. The emitter voltage should be zero, because there is no current flow through the base lead. Therefore, the entire V_{CC} supply voltage should be dropped between the emitter and collector leads.

If the circuit in question happens to be a common collector configuration, V_{CC} and the collector voltage would always be the same. Therefore, one final test can be performed to fully test the transistor for proper operation. Turn off the applied power to the circuit. With the base lead still disconnected from the previous tests, use

Fig. 14-16. Incorrect in-circuit transistor voltages.

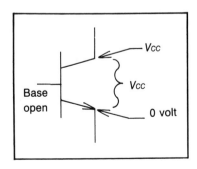

V_{CC}

V_{CC}

Base
open

0 volt

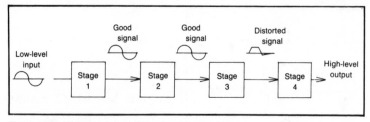

Fig. 14-17. Multistage transistor troubleshooting.

an ohmmeter to check for the proper diode junction indication between the base and collector. Since the base lead is open, the associated circuitry will not interfere with this check, and it is not necessary to disconnect the collector lead.

TROUBLESHOOTING TRANSISTORIZED CIRCUITS WITH AN OSCILLOSCOPE

An oscilloscope is a very valuable troubleshooting tool for observing the waveshape of a dynamic (moving) signal. For example, if you were experiencing a signal distortion problem involving many different transistor stages, an oscilloscope could be used to observe the waveshape from its point of origin through each consecutive transistor stage. The defective transistor stage would be the one in which the distortion first originated.

Figure 14-17 shows four transistorized stages. The first stage receives a low-level ac input signal for amplification. The amplified output of stage 1 is applied to the input of stage 2 for further amplification. This process is continued through all four stages with each stage providing additional amplification of the original signal.

An oscilloscope is used to observe the output of each stage (which is also the input to the succeeding stage). Since the waveshape from the output of stage 1 is undistorted (meaning it is the same waveshape pattern as applied to the input), the oscilloscope is used to observe the output of stage 2, which is also undistorted. When the output of stage 3 is observed, the oscilloscope shows this waveshape to be very distorted (unlike the original input waveshape). Therefore, the distortion must be occurring in stage 3.

Since you cannot observe a waveshape with a VOM or DVM, this type of problem could be very difficult to troubleshoot without an oscilloscope. Many schematic circuit diagrams include typical waveshape patterns at key test points within the circuit.

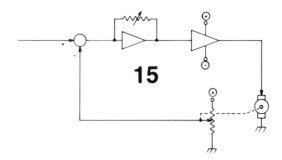

Thyristors

THE TERM *THYRISTOR* REFERS TO A BROAD FAMILY OF SEMI-conductor devices used primarily for power control. Thyristors are basically fast-acting electronic switches, including the SCR, the UJT, the triac, and the diac. All of these devices are used extensively in industrial electronic applications.

The *SCR (silicon-controlled rectifier)* is probably the most common of all thyristors. An SCR is a three-lead device that looks very much like a transistor. The three leads are the *gate*, the *cathode*, and the *anode*. An example of SCR construction is shown in Fig. 15-1, and the corresponding electrical symbol is shown in Fig. 15-2.

Like a diode, the SCR will allow current to flow in only one direction. The cathode must be negative in relation to the anode. But unlike a diode, a positive potential (or pulse) relative to the cathode must be applied to the gate before forward current flow can begin. Once the forward current flow begins, the SCR appears to be a short from cathode to anode that cannot be controlled by the gate.

One method of stopping the forward current flow is to reverse-bias the SCR (force the cathode to become positive relative to the anode). Another method is to allow the forward current flow to drop below the SCR's *holding current*. The holding current is a manufacturer's specification defining the minimum current required to hold the SCR in a conductive state.

If the forward current flow drops below the specified holding

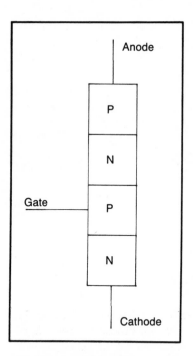

Fig. 15-1. Basic SCR construction.

current, the SCR will drop out of conduction. Once an SCR drops out of conduction (due to either a voltage polarity reversal or a loss of minimum holding current), control is again returned to the gate, and the SCR will not conduct (even if forward-biased) until another positive pulse (relative to the cathode) is applied to the gate.

Like the transistor, the SCR is considered a current device, be-

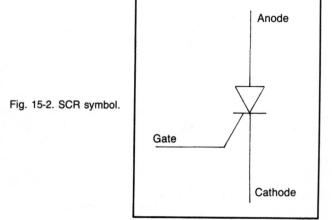

Fig. 15-2. SCR symbol.

cause the gate current causes the SCR to begin to conduct (if forward-biased). Also, the forward current flow (from cathode to anode) is what maintains conduction once the SCR is turned on by the gate current.

Because SCRs can be turned off when they are reverse-biased, they are very commonly used in ac power applications. Because ac power reverses polarity periodically, an SCR used in an ac circuit will automatically be reverse-biased (causing it to turn off) during one-half of each cycle. During the other half of each cycle, it will be forward biased but will not conduct unless a positive gate pulse (relative to the cathode) is applied. Because this positive gate pulse can be applied at any time during the half-cycle in which the SCR is forward-biased, the SCR can control the amount of given power to a load during that half-cycle.

Consider the circuit shown in Fig. 15-3. Had there been a method of closing switch S1 at the positive peak of the ac source, the SCR would be turned on and conduct because a positive pulse was received at the gate and the SCR was forward-biased.

The associated waveshapes of this circuit are illustrated in Fig. 15-4. During the beginning of the positive half-cycle (before S1 is closed), the SCR is not conducting; it appears to be an open switch. Therefore, the entire ac source voltage will be dropped across it.

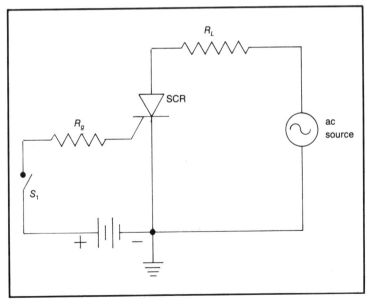

Fig. 15-3. Demonstration of SCR operating principles.

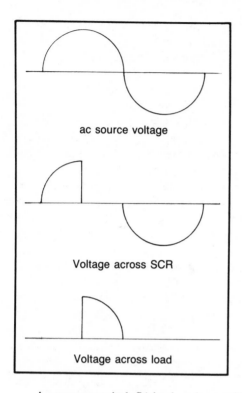

ac source voltage

Voltage across SCR

Voltage across load

Fig. 15-4. Waveshapes from circuit illustrated in Fig. 15-3.

As soon as switch S1 is closed, a positive pulse relative to the cathode is applied to the gate, causing the SCR to conduct and appear to be a short. This effect is similar to closing a switch between the ac source and the load (R_L). Therefore, the remaining portion of the positive half-cycle will appear across the load (R), and none of the remaining half-cycle will appear across the SCR.

As soon as the ac source reverses polarity into the negative half-cycle, the SCR is reverse-biased. This causes it to drop out of conduction, and it again looks like an open switch. This causes the entire negative half-cycle to appear across the SCR.

NOTE: Resistor R_g was placed in the gate circuit to limit the current so that the maximum allowable gate current is not exceeded.

You should understand several important points about the operation of this simple circuit. First, once the SCR has been turned on by closing the switch (S1), it cannot be turned off again during the remainder of the positive half-cycle. As long as it conducts current, the gate circuit loses all control. Secondly, before the SCR receives a gate trigger and begins to conduct, there is virtually no power consumption in the circuit. Once the SCR begins to conduct, virtually all of the power was delivered to the load.

Again, refer to Fig. 15-3. Assume the ac source was common 120 Vac household power. If you had a means of closing switch S1 at precisely this same point in time continuously (at the peak of the positive half-cycle, 60 times a second), you would actually be delivering 25 percent of the maximum available power to the load (R_L). By varying the point in time at which you closed the switch (S1), you could vary the power delivered to the load (R_L) from 0 percent to 50 percent. You could never exceed 50 percent, because the SCR cannot conduct during the negative half-cycle of the ac waveform.

The SCR wastes very little power while accomplishing this 0-to-50 percent power control, because it functions in either of two ways: "on" (analogous to a closed switch) or "off" (analogous to an open switch). A closed switch does not dissipate any appreciable power, because it does not present any opposition to current flow. Without any opposition to current flow, there cannot be any power dissipation, because a voltage drop is not developed $(P = IE)$. Nor does an open switch dissipate power, because it does not allow current to flow. Since virtually all of the power being drawn from the source is delivered to the load (R_L), the efficiency of this circuit would approach 100 percent throughout its entire range.

To understand the importance of efficient power control, consider another method of varying the power to a load. Figure 15-5 shows a circuit in which a rheostat is used to vary the power delivered to the 50-ohm load (R_L).

The rheostat is adjustable from 0 ohms to 50 ohms. When the rheostat is adjusted to 0 ohms, it will appear to be a short, and the entire 120-Vac source will appear across the load (R_L). The power dissipated by the load (R_L) can be calculated as follows:

$$P = \frac{E^2}{R} = \frac{(120)^2}{50} = \frac{14400}{50} = 288 \text{ watts}$$

Similarly, the power dissipated by the rheostat will be:

$$P = \frac{E^2}{R} = \frac{(0)^2}{0} = \frac{0}{0} = 0 \text{ watts}$$

If the rheostat is adjusted to present 50 ohms of resistance, the

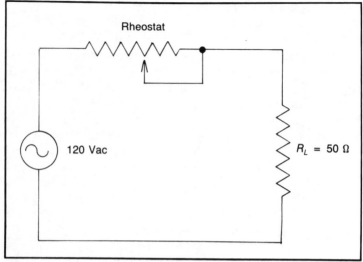

Fig. 15-5. Inefficient power control circuit.

voltage dropped by the rheostat will equal the voltage dropped by the load (R_L), because the resistance value of the rheostat and the load are equal. Therefore, 60 Vac will be dropped by both. Under these circumstances, the power dissipated by the load (R_L) will be:

$$P = \frac{E^2}{R} = \frac{(60)^2}{50} = \frac{3600}{50} = 72 \text{ watts}$$

The power dissipated by the rheostat will be:

$$P = \frac{E^2}{R} = \frac{(60)^2}{50} = \frac{3600}{50} = 72 \text{ watts}$$

Notice that both the rheostat and the load (R_L) dissipate the same amount of power. Since you only want to control the power to the load, the power dissipated by the rheostat is wasted. In this case, the efficiency of the power control is only 50 percent. Obviously, at different settings of the rheostat, different efficiency levels occur.

A disadvantage of using a single SCR for power control, as shown in Fig. 15-3, is that you cannot control both the negative and positive half-cycles of an ac waveform. To overcome this, two SCRs may be incorporated in a circuit for full-wave power control.

176

THE TRIAC

Another member of the thyristor family, the triac, can be used for full-wave power control. The triac has three leads: the *gate*, M_1, and M_2. A triac can be triggered by either a positive or negative pulse to the gate lead in reference to the M_1 terminal. It can conduct current in either direction between the M_1 and M_2 terminals. Like an SCR, once the triac has been triggered, the gate loses all control until the current flow between the M_1 and M_2 terminals drops below the manufacturer's specified holding current. Since a minimum specified gate current is required to trigger a triac, and a minimum specified holding current keeps the triac in a conductive state, it is considered a current device. The symbol for a triac is shown in Fig. 15-6.

The principle of efficient power control is essentially the same for the triac as it is for the SCR. Since a triac operates in only two modes (on or off), full-wave power control can be obtained without appreciable power losses in the triac itself. The primary advantage of the triac is its capability of being triggered in either polarity and controlling power throughout the entire ac cycle.

Triacs are commonly used for smaller power-control applications (light dimmers, small motors and power supplies). Unfortunately, triacs have the disadvantage of being difficult to turn off (especially when used to control inductive loads). For this reason, SCRs are used almost exclusively in high-power applications.

UJTS, DIACS, AND NEON TUBES

Until now, you have examined the theoretical possibility of con-

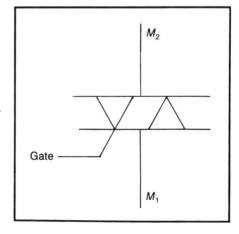

Fig. 15-6. Triac symbol.

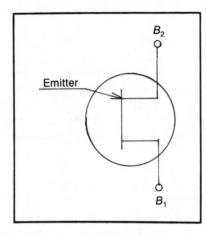

Fig. 15-7. UJT symbol.

trolling power with SCRs or triacs if you could vary the phase angle of the gate trigger pulse relative to the ac cycle. Obviously, it would be impossible for a human to turn a switch on and off at a 60-hertz rate and trigger an SCR or a triac at the same point in time during each cycle. The UJT, the diac, and the neon tube are commonly used to accomplish this function. (The neon tube is not actually a member of the thyristor family, but its function is identical to the diac. Some older industrial equipment may still use neon tubes for gate triggering.)

Like the SCR, the UJT (unijunction transistor) is a three-lead device. The three leads are B_1, B_2, and the emitter. The schematic symbol for the UJT is shown in Fig. 15-7. Unlike the previously discussed SCR and triac, the UJT is a voltage device. When the voltage between the emitter and B_1 lead reaches a certain value (a ratio of the applied voltage between the B_1 and B_2 leads and the manufactured characteristics of the device), the resistance between the emitter and B_1 decreases to a very low value. If the voltage between the emitter and B_1 decreases to a value below the established ratio, the resistance between the emitter and B_1 increases to a high value. In other words, you may think of a UJT as a voltage-breakdown device. It will avalanche into a highly conductive state (between the emitter and B_1 leads) when a peak voltage level (referred to as V_p) is reached. It will continue to remain highly conductive until the voltage is reduced to a much lower level called the *valley voltage* (V_v).

All UJTs can be incorporated into circuits which vary the amplitude of the applied ac voltage to the UJT and precisely control the V_p point relative to the ac applied to an SCR or triac. Utiliz-

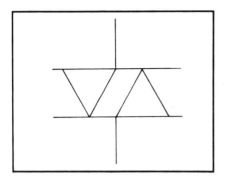

Fig. 15-8. Diac symbol.

ing the conductive breakdown characteristics of the UJT as a means of providing the trigger pulses, the SCR or triac can consequently be triggered repeatedly at any point in the ac cycle.

Another voltage breakdown device is the diac. The schematic symbol for a diac is illustrated in Fig. 15-8. The diac is a two-lead device very similar in physical appearance to a small diode. According to its manufacturing process, every diac has a preset breakover voltage point at which it will present a low resistance to current flow. Current may flow in either direction as long as the breakover voltage is obtained.

Once the breakover voltage has been reached (typically about 32 volts) and the diac begins to conduct, it will continue to conduct until the applied voltage across its two terminals drops to a lower level (typically 28 to 26 volts). As in the case of the UJT, once conduction begins, a current avalanche occurs providing a pulse suitable to trigger SCRs or triacs.

The neon tube is identical in operation to the diac except that it has a much higher breakover voltage (typically 80 to 90 volts).

A TYPICAL POWER-CONTROL CIRCUIT

Figure 15-9 shows a typical ac power-control circuit such as the type used in common incandescent light dimmers. The load (R_L) could be a light bulb or any ac-powered device. Note the load (R_L) is in series with the triac, and the gate trigger circuit (R_1, C_1, and the diac) is in parallel with the triac.

The following explanation will describe the circuit operation during one half-cycle of the applied ac. The polarity of the ac is irrelevant in this case, because both the triac and the diac will conduct in either direction.

If the instantaneous value of the applied ac voltage is zero, the triac is off, because the circuit current has dropped below the triac's

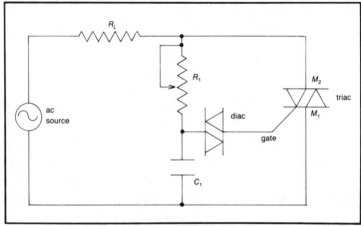

Fig. 15-9. Efficient triac power control circuit.

holding current. As the ac cycle continues and the applied voltage
increases, this entire voltage will be dropped across the triac. (The
triac appears to be an open switch at this point.) Likewise, the same
voltage is dropped across the trigger circuit, because it is in parallel
with the triac. At this point, C_1 will begin to charge at a rate relative to the setting of rheostat R_1. (The setting of rheostat R_1 actually varies the time constant of this circuit, as explained in Chapter
11.)

As the ac cycle continues, C_1 will eventually charge to the
specified breakover voltage of the diac, causing the diac to
avalanche and a current pulse (trigger pulse) to flow through the
gate and the M_1 terminal of the triac. This trigger pulse causes the
triac to turn on (much like a closed switch), and the remainder of
the ac half-cycle will be applied to the load (R_L).

As the ac cycle continues and approaches zero (prior to changing polarity), the circuit current drops below the triac's holding current, and it again returns to a nonconductive state. This entire
process repeats during the next half-cycle of the applied ac source.

There are several important points to understand about the
operation of this simple circuit. The diac will reach its breakover
voltage and trigger the triac at the same relative point during each
half-cycle of the ac waveform. This relative point will depend on
the charge rate of C_1, which is controlled by the setting of R_1.
Therefore, by varying the setting of R_1, you can control the percentage of each half-cycle applied to the load (R_L), which consequently varies the percentage of average power that it dissipates.

Throughout the entire power control range of this circuit, the power wasted by the triac will be negligible compared to the power applied to the load.

The simple circuit shown in Fig. 15-9 is intended to provide you with a basic understanding of thyristor operation. A more detailed discussion of thyristor operation will be covered in the succeeding chapters regarding dc and ac motor controls.

TROUBLESHOOTING THYRISTORS

Troubleshooting thyristors is more difficult than troubleshooting the previously discussed devices. In many cases, it may be easier to analyze the circuit problem and logically determine the component failure. Depending on the test equipment you have available, the following tests may help isolate the fault. It will be necessary to use one or more manufacturer's cross-reference manuals to determine the device's terminal designations based on its package style or part number.

To partially test an SCR with an ohmmeter, check for continuity between the anode and the cathode. You should see infinite resistance in both directions. The resistance between the gate and the cathode should be very low (but not zero) in both directions. Typically, an ohmmeter will measure about 20-100 ohms both ways. The resistance between the gate and the anode should be infinite in both directions.

Testing a triac with an ohmmeter is similar to testing an SCR. You should read a low resistance (not zero) with both ohmmeter polarities between the gate and M_1 terminals. As with the SCR, this resistance will typically be 20-100 ohms. You should read infinite resistance between the M_1 and M_2 leads and between the gate and M_2 leads.

Figure 15-10 illustrates a simple circuit for testing the operation of an SCR. The power source can be a battery (a 6-volt lantern battery works well for this purpose) or a dc power supply. Although the power supply voltage is not critical, it must not exceed the rated maximum cathode-to-anode voltage specified by the manufacturer, and it should not be less than 3 volts for a reliable test.

Depending on the power-source voltage, the limiting resistor is chosen so the maximum specified cathode-to-anode current is not exceeded. To calculate the value of this resistor, assume the SCR to be a closed switch, and use Ohm's law to solve for

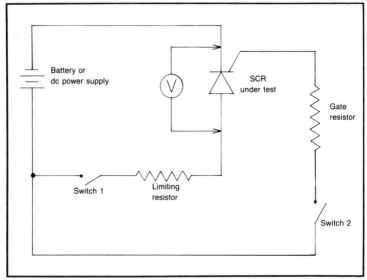

Fig. 15-10. Simple SCR test circuit.

resistance. (You know the power source voltage and the desired
cathode-to-anode current.) The value of the anode-to-cathode current is not critical as long as it exceeds the minimum holding current as specified by the manufacturer.

The gate resistor is chosen in the same manner as the limiting
resistor. It is calculated to limit the gate-to-cathode current below
the maximum specified gate current while allowing enough current flow to reliably turn on the SCR. The minimum gate current
for a reliable turn on is also specified by the manufacturer.

Once the circuit has been constructed, the following procedure
will test the dynamic operation of the SCR. A voltmeter is placed
across the SCR under test. When switch 1 is closed, the entire
power source voltage should appear across the SCR. As soon as
switch 2 is closed, the voltage across the SCR should drop to approximately 1 volt or less (this means the SCR is conducting current and the majority of the power source voltage will be dropped
by the limiting resistor). When switch 2 is opened, the SCR should
continue to conduct, and the voltage across it should remain at 1
volt or less. When switch 1 is opened, the voltage across the SCR
will obviously drop to zero, because the circuit has been opened.
When switch 1 is again closed, the entire power source voltage
should again appear across the SCR.

The previous test proves the SCR is capable of being turned

182

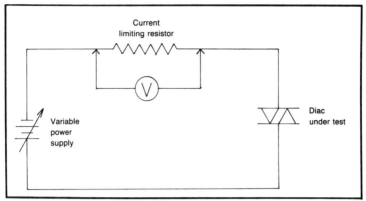

Fig. 15-11. Simple diac test circuit.

on by the gate and will continue to conduct cathode-to-anode current until this current falls below the minimum holding current, causing the SCR to turn off. Although this test appears complicated and time-consuming, in reality, it is rather quick and simple. The physical construction of the test circuit does not have to be elaborate. It can usually be fabricated with commonly available clip leads. Instead of installing physical switches, the clip leads can simply be connected and disconnected.

Dynamic testing of triacs is performed exactly the same as with SCRs. (See Fig. 15-10.) The triac gate is connected to the same point as the SCR gate, the M_1 terminal is connected to the same point as the SCR's cathode, and the M_2 terminal is connected to the same point as the SCR's anode. Under these conditions, a triac should operate exactly like an SCR. If it does, the leads to the power source should be reversed and the test repeated. This proves the triac is capable of conducting and being triggered in either polarity.

Diacs are most easily tested using a variable-voltage dc-power supply. A simple test circuit for this purpose is shown in Fig. 15-11. The current limiting resistor should be chosen to limit the diac current well below the manufacturer's maximum current specification. A voltmeter is placed across this limiting resistor, and the voltage output of the variable voltage power supply is slowly increased (from zero) while the voltmeter is observed. At a voltage output below the breakover voltage of the diac, no voltage should be dropped by the current-limiting resistor.

When the breakover voltage of the diac is obtained, you should observe an abrupt increase in voltage across the resistor and, as you continue to increase the power supply voltage, the increase

183

should appear proportionally across the resistor.

It is usually easier to test UJTs dynamically while the circuit is in operation. An oscilloscope is needed for this purpose. Chapter 25, which deals with dc motor controls, explains this subject in greater detail.

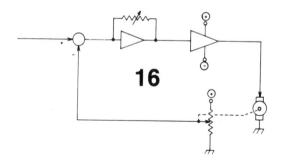

Other Discrete Semiconductors

THE TERM *DISCRETE SEMICONDUCTOR* REFERS TO A NONIN-tegrated semiconductor. Integrated circuits will be discussed in a later chapter.

THE FIELD-EFFECT TRANSISTOR (FET)

The *field-effect transistor* (FET) is an active-voltage device. Unlike the bipolar transistor, an FET is not a current amplifier. It is much like a vacuum tube in basic operation.

An FET is a three-lead device similar in physical appearance to a transistor. The three leads are referred to as the *source*, the *drain*, and the *gate*. These leads are somewhat analogous to the junction transistor's emitter, collector, and base leads respectively. There are two general types of FET: the *junction field-effect transistor (JFET)* and the *insulated-gate metal-oxide semiconductor field-effect transistor* (MOSFET or IGFET).

FETs are manufactured as either N-channel or P-channel types. N-channel FETs are used in applications requiring the drain to be positive relative to the source. The opposite is true of P-channel FETs. The schematic symbols for N-channel and P-channel JFETs and MOSFETs are shown in Fig. 16-1. The arrows always point toward the channel (the interconnection between the source and drain) in N-channel devices and away from it in P-channel devices.

All FETs have one thing in common—a very high gate-input

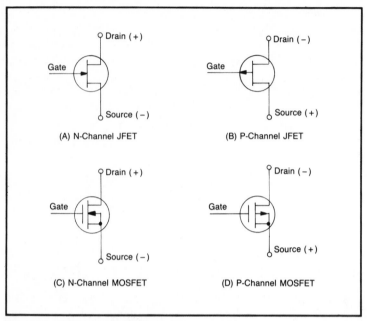

Fig. 16-1. JFET and MOSFET symbols.

impedance (1 megohm to over 1,000,000 megohms). This is due
to the fact that an FET is a voltage device. The gate resistance
(impedance) is equal to the gate voltage divided by the gate current (Ohm's law). Since the gate does not need current to function,
FETs can be manufactured to present an almost infinite input impedance. This characteristic can be very useful if you do not want
to load a preceding circuit or input. (Loading was discussed in Chapter 14.) The input impedance is much higher in MOSFETs because
the gate is actually insulated from the device by a very thin oxide
layer.

There are many different biasing arrangements for FETs, depending upon the application, but the basic operating principle is
the same for all of them. A voltage applied to the gate will produce
an electrostatic field which will increase or decrease a depletion
region between the source and drain. This causes a respective decrease or increase in current between the source and drain. You
may think of an FET as a voltage-controlled resistor. The increase
or decrease in source-to-drain current may be converted to a respective voltage drop (and gain) in the same way as in standard bipolar
transistor circuits.

As previously discussed, the primary-gain parameter of a standard bipolar transistor is beta. Beta defines the ratio of the current flow through the base relative to the current flow through the collector. When considering FETs, the primary-gain parameter is called the *transconductance* (G_{f_s} or G_m). The transconductance is equal to the ratio of the change in drain current to the change in gate-to-source voltage. The transconductance is typically defined in terms of micromhos (the *mho* is the basic unit for expressing conductance). Typical transconductance values for common FETs range from 2,000 to 15,000 micromhos. The equation for calculating transconductance is:

$$G_{f_s} = \frac{\Delta \, I_D \text{ (change in drain current)}}{\Delta \, V_{GS} \text{ (change in gate-to-source voltage)}}$$

For example, if in a particular FET a 1-volt change in the gate-to-source voltage caused a 10-milliamp change in the drain current, the transconductance of that FET would be:

$$G_{f_s} = \frac{\Delta \, I_D}{\Delta \, V_{GS}} = \frac{10 \text{ mA}}{1 \text{ volt}} = 10 \times 10^{-3} = 10,000 \times 10^{-6}$$

$$= 10,000 \text{ micromhos}$$

Refer to Fig. 16-2. Assume the FET in this circuit has the same transconductance characteristics as shown in the previous example and that the value of the drain resistor (R_d) is 1 kohm. A 1-volt change on the input (causing a corresponding 1-volt change in the gate-to-source voltage) would cause a 10-milliamp change in the drain current. According to Ohm's law, a 10-milliamp change through the 1-kohm drain resistor will cause a 10-volt change across the drain resistor. This 10-volt change will appear at the output. Therefore, since a 1-volt change on the gate caused a 10-volt change on the output, this circuit would have a voltage gain of 10.

In several ways, FET circuits can be compared with standard bipolar transistor circuits. The circuit shown in Fig. 16-2 is analogous to the common emitter circuit, and is appropriately called a *common source configuration*. The output is inverted (180 degrees out of phase) from the input, and it is capable of voltage and current gain. If the output were taken from the source instead of the drain, it would then be a common drain configuration. The output would not be inverted, and the voltage gain would be approximately 1. This configuration is analogous to the common collector bipolar transistor circuit.

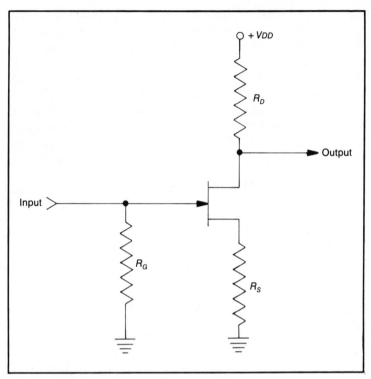

Fig. 16-2. Common source configuration JFET circuit.

FET BIASING CONSIDERATIONS

With N-channel JFETs, the gate will always be biased negative relative to the source. Likewise, P-channel JFETs require a positive gate voltage relative to the source. A JFET exhibits maximum conductivity (minimum resistance) from the source to the drain with no bias voltage applied to the gate.

Refer again to the circuit shown in Fig. 16-2. Note that the gate is effectively placed at the same potential as circuit common through resistor R_g. With no input applied, the gate voltage (relative to circuit common) is zero. This does not mean the gate-to-source voltage is zero. Assume that the source resistor (R_s) is 100 ohms and that the drain current (which is the same as the source current) is 10 milliamps. These conditions would cause a 1-volt drop across the source resistor (R_s), putting the source at a positive 1-volt level above the circuit common. If the source is 1 volt more positive than the gate, you could also say the gate is 1 volt negative relative to the source. Therefore, the gate-to-source voltage in this case is − 1

188

volt. If an input was applied to this circuit, causing the gate-to-source voltage to increase to -2 volts, the FET would become less conductive (more resistive) between the source and the drain, causing the drain current to decrease.

A common method of biasing a MOSFET is shown in Fig. 16-3. Depending on how it is manufactured, the gate of an N-channel MOSFET may be either negative or positive relative to the source.

As in the case of N-channel JFETs, as the gate becomes more negative relative to the source, the drain current is reduced. Unlike an N-channel JFET, as the gate voltage becomes more positive relative to the source, the drain current is increased.

A P-channel MOSFET operates identically, except that the voltage polarities are reversed. In this circuit, the gate-to-source bias voltage will be determined by the voltage divider network of R_1 and R_2 and the value of R_s relative to the drain current.

As in the case of bipolar transistors, there are many other types

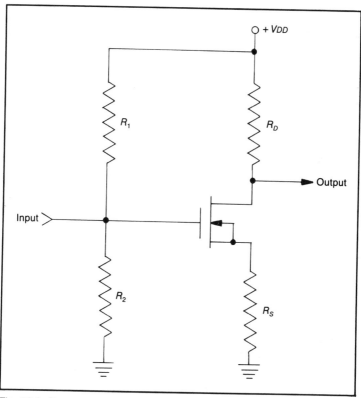

Fig. 16-3. Common source configuration MOSFET circuit.

of FET circuits and biasing arrangements. If the circuits illustrated in Figs. 16-2 and 16-3 are used for amplifying ac signals, the inputs and outputs may be capacitor-coupled to block any dc voltages from interconnecting stages.

TESTING FETS

Caution should be exercised when testing any MOSFET device.

Because of the extremely high input impedance, even a small static voltage on your body (too small for you to detect) can destroy the device. Before touching any MOS device, ground your body, and then frequently continue to ground your body during servicing of the equipment. If you anticipate continued servicing of MOS devices, you may want to invest in special static-protection wrist straps or mats available from many companies.

As in the case of bipolar transistors and thyristors, a good selection of manufacturer's cross reference manuals will be needed for determining the lead designations of FETs. If a defective FET is located, these manuals will also be useful in finding the correct replacement FET.

The testing of FETs out-of-circuit cannot be performed reliably with an ohmmeter alone. If you must test FETs out-of-circuit, you should purchase a transistor tester. A good transistor tester should be able to test JFETs, MOSFETs, and bipolar transistors.

It is much more desirable to test MOSFETs and JFETs in-circuit under dynamic operating conditions. This is not difficult to accomplish with a little common sense and a basic conceptual knowledge of basic FET operation. For example, if you measured a positive gate-to-source voltage on an N-channel JFET, either the JFET is defective or something is wrong with the biasing circuit. If you observed a varying gate voltage on a MOSFET without any consequential change in the source or drain voltages, you should suspect a defective MOSFET or other associated circuit component.

As with bipolar transistors (see Chapter 14), an oscilloscope can be very useful in troubleshooting multistage FET circuits. The procedure would be the same as that described for bipolar transistors.

PHOTOCONDUCTIVE CELLS

A *photoconductive cell*, or *photocell*, is a two-terminal resistive device. Its resistance changes if the intensity of light shining on it changes. In absence of light, the cell resistance is maximum. This resistance typically drops to less than a tenth of its maximum when the cell is exposed to light. These cells can be checked easily by measuring the cell resistance with an ohmmeter while covering the lens and then shining a flashlight on the lens and checking for an appropriate change in resistance.

PHOTODIODES AND PHOTOTRANSISTORS

A *photodiode* is a two-terminal device. It is specially constructed so that its leakage characteristics change drastically when the device is exposed to light. Typically, a photodiode is designed to be used in a reverse-biased circuit configuration. When not exposed to light, its conductivity is very low. When exposed to light, its conductivity changes in respect to light intensity.

The *phototransistor* has an optical lens which allows the base-to-emitter junction to be exposed to light. In a properly designed circuit, the collector current can be increased by increasing the intensity of light falling on the lens assembly.

You can usually check these devices for proper operation by using a flashlight and noting the appropriate change.

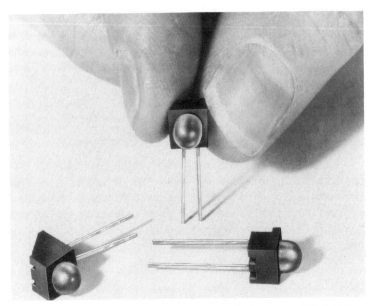

An example of the relative size of some typical LEDs (courtesy of Dialight Corporation, a North American Philips Corp.).

A 4-array LED combination for displaying logic conditions (courtesy of Dialight Corporation, a North American Philips Corp.).

LIGHT-EMITTING DIODE

The *light-emitting diode* (*LED*) is a specially designed diode that emits light when forward-biased. This light is usually in the visible spectrum, but it may be infrared depending upon the application.

An LED is usually used as an indicator. It can also be used with photo devices to sense the presence or absence of objects for counting or controlling applications. Like any other standard diode, an LED may be checked with an ohmmeter.

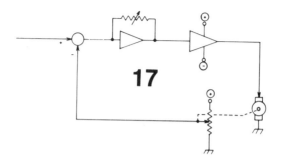

Integrated Circuits

THE PROCESS OF MINIATURIZING MULTIDEVICE CIRCUITS (transistors, resistors, and diodes) is called *integration*. An *integrated circuit* (*IC*) is a chip that contains (or can do the function of) many discrete devices. In appearance, the most common type of IC is a small, rectangular package with 14 or 16 pins (or legs) extending from the package. This package style is commonly called a *dual in-line package* or *DIP*. Other common package styles include round metal casings with multiple leads extending from them and larger rectangular packages with up to 40 pins.

An integrated circuit is designed for a specific function. For example, a LM310AH is specifically designed to be used as an operational amplifier.

INTEGRATED CIRCUITS IN GENERAL

There are two main classifications of ICs: *digital* and *analog*.

Digital ICs are designed for use in digital applications—that is, applications requiring the processing of only logic levels (lows and highs). Small-scale integrated circuits (SSI) and medium-scale integrated circuits (MSI) can perform logic functions, such as OR, AND, NOR, NAND, NOT, EXCLUSIVE OR, addition, subtraction, up-down counting, and latching. (Digital functions will be discussed in more detail later in this chapter.) Large-scale integrated circuits (LSI) can be manufactured as complete microprocessors and other complex devices.

Analog ICs, sometimes referred to as linear ICs, are designed to process infinitely variable signals, such as audio signals and process-feedback voltages.

There are four primary uses for analog ICs, which you are likely to encounter in industrial electronics. These include amplification, regulation, comparison, and filtering.

Ampification refers to the process of increasing a signal voltage or current to a higher level. Amplification is analogous to gain, which has been discussed in previous chapters.

Regulation refers to the process of holding a voltage or current at a constant level regardless of load variations. *Regulators* are commonly used in power supplies. A *zener diode* (discussed in Chapter 13) is an example of a discrete voltage regulator. Among the advantages of IC regulators is that they can provide a much higher degree of regulation.

An analog circuit designed to compare the higher of two analog signals is called a *comparator*. A comparator produces an output whenever an analog signal voltage goes above a specified limit or threshold.

Filtering refers to a certain frequency or group of frequencies being subtracted from an ac voltage. *Integration* and *differentiation* are two forms of filtering. Integration is high-frequency filtering, and differentiation is low-frequency filtering. Electronic filtering is analogous to filtering a mixture of large and small particles. For example, if you wanted to filter out all of the large stones from a

A few examples of the many IC package styles in current use (courtesy of Fairchild Semiconductor Corporation).

mound of sand, you could sift the sand through a screen, which would block all of the larger stones from passing through. Likewise, an integrator will filter out high-frequency ac waveshapes and only allow the low frequencies to pass freely. In contrast, a differentiator will block lower frequencies and allow the high frequencies to pass.

GENERAL IC TROUBLESHOOTING PROCEDURES

NOTE: You must exercise caution when working with some of the newer types of ICs. The MOSFET input operational amplifiers and the CMOS digital ICs are susceptible to destruction by static discharges.

When troubleshooting ICs, the following steps should be taken:

1. Obtain the generic device number from the IC or manufacturer.
2. Find the operational specifications from an appropriate manufacturer's cross-reference manual, or call a local electronics dealership for the device specifications.
3. Verify that the proper in-circuit operating parameters are being met by examining the IC dynamically. In essence, this means you should measure the operating voltages and inputs to the IC in question.
4. If all of the operating parameters are correct and the IC is still not functioning properly, replace it.
5. Test to verify the replacement IC is functioning properly.

OPERATIONAL AMPLIFIERS

An *operational amplifier* (*op-amp*) is a two-input, high-gain, basic amplifier package. One IC chip may contain as many as four operational amplifiers, but most contain only one. The schematic symbol for an operational amplifier is shown in Fig. 17-1.

Operational amplifiers will almost always operate from a dual power supply (a power supply with both a positive and negative output). Typical voltage ranges are from 5 Vdc to 18 Vdc. Op-amps are typically low-power output devices, so power consumptions are also low (500 mW or less). All operational amplifiers within the same IC will operate from the same power-supply input pins.

The two-signal inputs for each op-amp are the *inverting* and

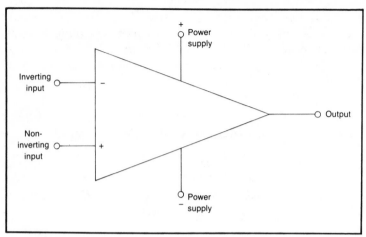

Fig. 17-1. Operational amplifier symbol.

non-inverting inputs. An input signal applied to the inverting input will be inverted (or 180 degrees out of phase) at the output. The converse is true for the non-inverting input. (The output will be in phase.)

The voltage gain (open-loop gain) of an op-amp is very high (1×10 to the 6th power or higher). This high gain is useful for comparator applications but is not practical for signal-amplification applications. For this reason, the design engineer must usually incorporate some negative feedback to limit the gain to a reasonable value. This is performed easily by placing a resistance from the output of the op-amp to the inverting input. In this closed-loop configuration, the voltage gain becomes a ratio of the feedback resistance to the input resistance.

A simple circuit providing negative feedback is shown in Fig. 17-2. The voltage gain of this circuit is approximately equal to the value of R_2 divided by the value of R_1. For example, if R_2 is 10 kohms and R_1 is 1 kohm, the voltage gain of this circuit would be approximately 10.

For filtering or nonlinear applications, the feedback may be inductive, capacitive, or something other than purely resistive. In some cases, the feedback may simply be a short from the output to the inverting input with the signal input applied to the noninverting input, as shown in Fig. 17-3. The op-amp then becomes a buffer. The voltage gain is one, but the input impedance is very high and the output impedance is very low. This buffer circuit is useful for matching a high-impedance output to a low-impedance

196

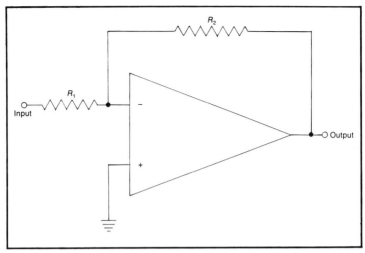

Fig. 17-2. Operational amplifier configured as a voltage amplifier.

input. Op-amps may also be configured to perform arithmetic functions, such as square root extraction, exponential resolutions, addition, and subtraction. Unique circuits, such as oscillators, sample-and-hold, and even digital interfaces may be constructed. Some manufacturer's data books (available from your local electronics dealership or from the manufacturer), provide the schematics of many op-amp circuits for various applications.

Because of the inverting and non-inverting input configuration of an op-amp, it has a unique characteristic called *common mode rejection*. Common mode rejection simply means that if an input sig-

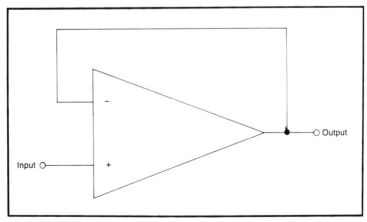

Fig. 17-3. Operational amplifier configured as a voltage follower.

nal of the exact waveshape and amplitude is applied to both inputs simultaneously, it will be rejected. This is because the inverting input will try to output the exact opposite of the non-inverting input, and the two will cancel one another. Common mode rejection is extremely useful in reducing electrical noise in long transmission lines.

As you may have realized by now, op-amps are extremely versatile. But don't let this versatility confuse your perception of these devices. They are still basically high-gain amplifiers. You may analyze their function within a circuit by analyzing the external components connected to them.

TROUBLESHOOTING OPERATIONAL AMPLIFIERS

Isolating circuit malfunctions to an op-amp fault can be accomplished in several ways. One way is to analyze the circuit and isolate the fault by reasonable deduction. Another way is to fabricate a test probe to simulate inputs and observe the op-amp's response to these inputs. This test probe is simply a 1-kohm resistor with one end connected to the positive or negative power supply used to power the IC. (The connection can be made easily with an alligator clip-lead or similar temporary connection device.) The other end of the resistor is used to touch the signal inputs of the op-amp while monitoring the output. (The resistor is necessary to limit current that could damage some external discrete components associated with the op-amp under test.)

When using the test probe, you can expect the following results:

1. If one end of the probe is connected to the positive supply:

   a. The output will go positive if the other end of the probe is touched to the non-inverting input.
   b. The output will go negative if the other end of the probe is touched to the inverting input.

2. If one end of the probe is connected to the negative supply:

   a. The output will go negative if the other end of the probe is touched to the non-inverting input.
   b. The output will go positive if the other end of the probe is touched to the inverting input.

It is important to realize that, although the above results can be expected, they are not absolute. In other words, depending on how the op-amp is configured in the circuit, the actual voltage output levels obtained will vary drastically. The important element to look for is the direction of change. If the direction of change is correct, you can be reasonably sure the op-amp is good.

There may also be external discrete components which either limit or inhibit the expected response. For example, if the output is connected with a diode to the ground to keep it from going negative, obviously the output will not go negative during the test, either. In that case, one end of the diode might have to be disconnected temporarily to properly test the op-amp.

To use another example, the same diode may be a zener diode intended to limit the maximum output to 5 volts. In this case, the output could never exceed 5 volts during the test, either.

If the op-amp under test happened to be configured as a buffer (see Fig. 17-3), touching the test probe to the inverting input would not affect the output. In this case, the op-amp could be tested by simply touching the test probe to the non-inverting input and verify the output follows the input precisely.

DIGITAL CONCEPTS

The term *digital* refers to devices or systems which perform logical functions utilizing a base-2 numbering system called the *binary* system.

The binary system only uses two digits—1 and 0. This is not as complicated as it may seem. Consider the decimal numbering system you are already familiar with. The term *decimal* means *base 10.* If you break down the number 1,543 into weights, you come up with 3 units, 4 tens, 5 hundreds, and 1 thousand. Notice how each succeeding weight is actually 10 times the value of the preceding weight. In other words, 1×10 is 10, 10×10 is 100, 100×10 is 1000, etc.

The binary numbering system works the same way except that it is based on two instead of ten. For example, the first weight (or least significant digit) is the units column. The next weight is 2×1, or 2. The next weight is 2×2, or 4. The next weight is 2×4, or 8. Instead of the number weights being units, tens, hundreds, thousands, ten-thousands, etc., the binary weights will be units, twos, fours, eights, sixteens, etc.

In the decimal system, there are ten possible numbers which

can be placed in any weight position (0, 1, 2, 3, 4, 5, 6, 7, 8, 9). In the binary system, there are only two possible numbers for any one weight position (0, 1). If you break down a binary number such as 0111 into weights, it means you have 1 unit, 1 two, 1 four, and 0 eights. By adding the weights together, you can convert a binary number into a decimal. In the previous example, $1 + 2 + 4 = 7$. Therefore, the decimal equivalent to 0111 is 7.

The following example demonstrates how you could count up to 9 using the binary numbering system:

0000 = 0	0101 = 5
0001 = 1	0110 = 6
0010 = 2	0111 = 7
0011 = 3	1000 = 8
0100 = 4	1001 = 9

The binary numbering system is convenient to use with electronic devices, because the binary digits 1 and 0 can be represented by an electronic device being either on or off. Binary digits can also be represented by either high or low non-absolute voltage levels. The only important factor is whether or not the voltage level is above or below an established reference level. For example, if you choose 1 volt as the reference level, any voltage less than 1 volt would be considered a *logical low,* or 0. Any voltage level above 1 volt, would be considered a *logical high*, or 1.

NOTE: Digital concepts will be discussed in greater detail in Chapter 18.

LOGIC GATES

Figure 17-4 illustrates a few of the more common logic gates you will probably encounter in IC form. Although there are many other types of logic gates and devices, you can always determine their logical function by analyzing their corresponding truth tables. The logic device, operating parameters, logic symbol, and truth table can be found in the manufacturer's data books or cross-reference manuals.

Refer to the AND gate shown in Fig. 17-4. Its corresponding truth table is shown in Fig. 17-5. This truth table states that the output of the AND gate will not go high (logical 1) until the A input and the B input also go high (logical 1). As long as either the

A or the B input is low (logical 0), the output will remain low.

Note the difference in the logical function of the OR gate shown in Fig. 17-4 and its truth table shown in Fig. 17-6. The output of an OR gate will go to a logical 1 whenever the A input or the B input goes to a logical 1. The NOT gate shown in Fig. 17-4 simply outputs the complement (or opposite) of whatever appears on its input. The truth table shown in Fig. 17-7 shows how the NOT function can be defined logically.

If a NOT gate is placed on the output of an AND gate, the gate become a NAND gate (NAND is the abbreviation for NOT AND). Likewise, if a NOT function is placed on the output of an OR gate, the gate becomes a NOR gate. The NOT (complement) function is indicated by the placement of a small circle on the output of the gate. This is shown on the 3-input NAND gate in Fig. 17-4.

The truth table for the 3-input NAND gate is shown in Fig. 17-8. Notice that the output is at a logical 1 unless all three inputs are at a logical 1. When all three inputs go to a logical 1, the output goes to a logical 0. If this were a 3-input AND gate, the output would stay at a logical 0 unless all three inputs were at a logical 1. In this case, the output would also go to a logical 1.

The NOT function can also be indicated on logic-gate inputs with a small circle just as it is indicated on gate outputs. The AND, NAND, OR, and NOR gates are available with as many as eight inputs per gate.

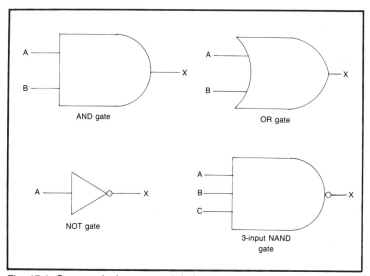

Fig. 17-4. Common logic gates and their associated symbols.

A	B	X
0	0	0
0	1	0
1	0	0
1	1	1

2-input AND gate
truth table

Fig. 17-5. Two-input
AND gate truth table.

Fig. 17-6. Two-input
OR gate truth table.

A	B	X
0	0	0
0	1	1
1	0	1
1	1	1

2-input OR gate
truth table

A	X
0	1
1	0

NOT gate
truth table

Fig. 17-7. NOT gate truth table.

Besides the simple logic-gate functions examined previously, there are literally thousands of different types of digital integrated circuits currently available. Digital ICs count, store, convert, and multiplex data, and provide clock pulses for timing purposes (oscillators).

TROUBLESHOOTING DIGITAL INTEGRATED CIRCUITS

CAUTION: Certain families of digital ICs (such as CMOS) are susceptible to destruction by static electricity. Be-

A	B	C	X
0	0	0	1
0	0	1	1
0	1	0	1
0	1	1	1
1	0	0	1
1	0	1	1
1	1	0	1
1	1	1	0

3-input NAND
gate truth table

Fig. 17-8. Three-input NAND gate truth table.

**fore you work with these devices, verify that the proper
safety precautions are taken.**

Although this chapter cannot include all of the information you
need to become well acquainted with the digital field, many good
books are available on the subject. Fortunately, for troubleshooting purposes, virtually all of the information you need is included
in the manufacturer's data books and cross-reference manuals.

There are some variations in troubleshooting digital ICs as compared to analog ICs. The IC in question should be located in the
manufacturer's data book or manufacturer's cross-reference manual to identify the chip's logical function, operating parameters, and
pin identification.

The logical function of the chip, for all possible input conditions, is described in the corresponding truth table. With this information, the troubleshooter may check the inputs and verify that
the proper outputs exist. This method of troubleshooting is most
easily performed during actual system operation.

If it is impossible to troubleshoot the IC in question during system operation, a way must be found to apply power to the printed
circuit board containing the IC and artificially inject an input (or
set of input conditions) to verify that the correct output occurs. Caution should be exercised when troubleshooting in this manner. The
inputs to one IC may be directly connected to the output of another
IC. This can cause interaction problems and possible destruction

of outputs. The schematic of the printed circuit board should be examined carefully before any attempt is made to artificially apply logic levels to the pins of any IC chips. It may be necessary to isolate certain pins on the IC in question before testing it.

When an IC is plugged into an IC socket, it may be carefully removed and placed in a temporary test circuit. There are many types of convenient breadboard systems available that allow you to fabricate custom test circuits in a matter of minutes. A good breadboard system is well worth the investment if you need to do this often. Before designing a custom-test circuit to test a specific IC, you should consider the cost-effectiveness of this decision. The majority of digital ICs are very inexpensive. When you consider the value of your time, it may be more cost-effective to simply replace the IC in question.

When troubleshooting large, complex microprocessor systems, you may not be able to isolate a problem to a component level due to the complexity of the inputs and timings involved. In these systems, problems are usually isolated to a board level, and the defective printed circuit board can be replaced. The defective board is then sent back to the manufacturer or a company specializing in industrial electronic repair of such items. But, it may be cost-effective for you to check what you can on these boards before sending them out for repair. Often, the problem is due to corrosion, shorted capacitors, defective transistors, or easily-tested ICs.

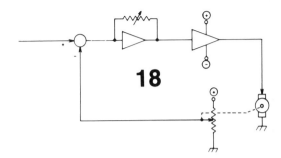

Digital Concepts and Systems

THIS CHAPTER IS DESIGNED TO ACQUAINT YOU WITH SOME OF the basic concepts of digital circuits and systems. If you have not previously encountered digital systems, some of the devices and concepts in this chapter will probably have little meaning for you. In later chapters, you will begin to understand how these building blocks work together to form a complete industrial control system, and you will probably want to refer back to this chapter from time to time. Don't become discouraged if you are confused while reading certain sections of this chapter for the first time.

Digital circuits are sometimes called *logic circuits* because they perform preprogrammed logical functions. A simple logic circuit provides an output that depends upon an actual set of input conditions. This type of circuit is, therefore, referred to as a *conditional logic circuit*. A more complex circuit can remember what has happened in previous steps and will provide an output reflecting both input conditions and the particular step in which it happens to be. This type of logic circuit is referred to as *sequential*. The most complex logic circuits employ computers or microprocessors to perform complex logic functions. These are referred to as *logic systems* and will be discussed later in this chapter.

Conditional and sequential logic circuits are simply a number of relays and contacts providing either open or closed circuit conditions. *Relay logic circuits* have been around for many years and are still providing reliable operation for industrial control systems.

Complex relay logic systems will utilize a relay ladder diagram to easily show the series of logical events and outputs that should occur. In some cases, a sequential relay logic circuit will use a stepper relay (a relay which incrementally steps through a series of multiple contacts) to increment control steps. In other cases, it will employ a cam controller (a cylindrical rotating drum with preset multiple contacts) for its memory. (Stepper relays and cam controllers were used in older clothes washers and dryers to control the preset washing and drying cycles.)

TROUBLESHOOTING RELAY LOGIC CIRCUITS

Under normal circumstances, the only test equipment needed for troubleshooting relay logic circuits is a VOM or DVM. If a relay ladder diagram, timing diagram, or schematic is available, you should carefully examine this documentation and fully understand the desired operation before proceeding. If the circuit documentation is not available, you may have to spend considerable time analyzing the circuit operation.

The majority of problems encountered with relay logic circuits are defective internal relay contacts. Depending on what the relay contacts are controlling, they could be exposed to electrical arcing, which can cause the contacts to eventually pit or deteriorate. Relay contacts can often be checked with power applied to the circuit. A good closed contact should not show any voltage drop across it. If the contact is open, a voltage will usually be read across it. (Never try to read resistance of a relay contact with power applied to the circuit.) In sequential circuits, remember to check the cam controller or stepper relay (and their associated switches) for proper operation.

Once again, a good visual examination is very important. Broken wires or loose connections make up a large percentage of problems associated with these circuits. The constant mechanical vibration from the relay operation itself can be severe enough to cause malfunctions.

OTHER DIGITAL CIRCUITS

Digital circuits are also manufactured using discrete semiconductors to replace electromechanical relays. The basic logic is the same, but instead of open or closed contacts, these circuits use high- or low-voltage levels to perform the same functions.

Most modern digital circuits use the integrated circuit equiva-

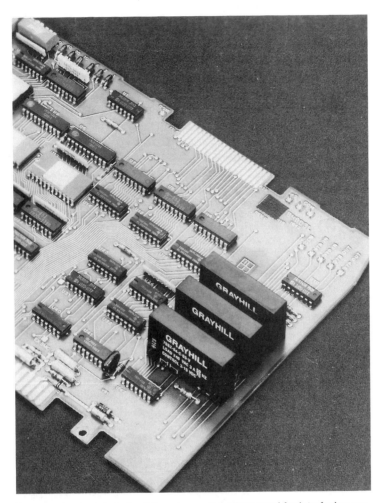

An example of printed circuit board mounted relays used for interfacing purposes (courtesy of Grayhill Inc., LaGrange, IL).

lent to relay logic called *digital* or *logic ICs*. These circuits are more reliable, use less power, are smaller, and are more versatile because of the numerous logic functions available. These circuits are also much less expensive. Digital ICs are manufactured according to a specific logic function. For example, an SN7400 IC contains four 2-input NAND gates. Common functions available in IC form are AND, OR, NOT, NAND, NOR, EXCLUSIVE OR, flip-flops, up-down counters, and decoders. For sequential circuits that are time-based, various forms of oscillators (clocks) are also available.

TROUBLESHOOTING SOLID-STATE LOGIC CIRCUITS

Troubleshooting solid-state logic circuits is essentially the same for IC or discrete forms. After a thorough visual examination, you should become acquainted with the logical operation of the circuit. If good circuit documentation is not available, the manufacturer may have to be contacted. Once this has been completed, the circuit operation can be observed, and the problem can be isolated to a specific component.

Troubleshooting digital ICs is discussed in the previous chapter.

LOGIC SYSTEMS

Whenever numerous logic circuits and gates are tied together for a more complex logical function, the resulting structure is called a *logic system*. Besides being able to simply turn devices on and off, logic systems are used extensively for interpreting and controlling analog (continuously variable) signals. Logic systems are capable of precise control of practically any industrial process. The size and complexity of logic systems can vary from only a few combinational/sequential circuits to massive multi-computer controlled plant-wide systems.

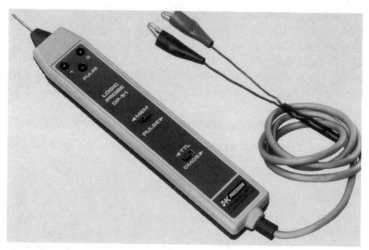

A digital logic probe used to check the logical status of several IC logic families (courtesy of B & K-Precision/Dynascan Corp.).

Combinational/Sequential Building Blocks

By combining the basic logic gates previously discussed (AND,

OR, NOT, FLIP-FLOP, etc.), some basic building blocks can be constructed and used universally in all data-processing and data-acquisition applications.

Data Selectors (Multiplexers)

In virtually all logic systems, there is a need to select one of many inputs for processing and analysis. In a *data selector*, multiple inputs are applied simultaneously while several control lines choose which input will be applied to the single output line. This is necessary, because microprocessors and computers can only see one input at a time. (The rapid rate at which examination occurs makes it appear immediately for all inputs.)

Data Routers (Demultiplexers)

Just the opposite of a data selector is the *data router*. A single output from a microprocessor or computer can be routed to multiple output lines by means of the data router. As in the case of the data selector, the control lines will decide which output line the information is routed to.

Encoders/Decoders

Encoders and *decoders* are used to convert data from one format to another. For example, an encoder is needed to convert decimal data from an input device to binary data for interpretation by a microprocessor or computer. Likewise, a binary output from the microprocessor or computer must go into a decoder for conversion to decimal data for human interpretation. Encoders are always used to convert some numbering system (other than binary) into binary. Decoders convert binary to any other numbering system.

Interfaces

In some cases, it is necessary to process or hold binary information in some manner for proper access to or from a microprocessor or computer. *Interfaces* are used for this purpose. For example, serial data (digital information transmitted on a single line) is usually converted to parallel data (digital information appearing simultaneously on multiple-data lines) for storage or processing. In this case, a serial to parallel interface is needed. To transmit parallel data long distances, a parallel to serial interface is used. In some cases, a buffer, or temporary storage area, is needed to correct any differences in feed-rate of data from one device to another. Sometimes, interfaces are also used for conversion from one logic family to another. (For example, TTL to CMOS.)

Counters

Digital counters have multiple applications in modern control systems. In simple sequential systems, the counter may actually take the place of the drum controller. The counter is also a basic building block in stepper motor drives and servo drives with encoder feedback.

THE COMPUTER

The *computer* is a digital logic system. A computer consists of four basic parts:

- The input section.
- The CPU (central processing unit).
- The memory.
- The output.

In order for a digital system to be called a computer, it must meet five essential criteria:

1. It must have input capability.
2. It must have memory to store data.
3. It must be capable of making calculations.
4. It must be capable of making decisions.
5. It must have output capability.

HOW A COMPUTER WORKS

A block diagram of a computer is shown in Fig. 18-1. The *input section* accepts information from a selected input device and converts it into digital information, which can be understood by the *central processing unit*. The CPU controls the timings and data selection points involved with accepting inputs and providing outputs by means of the *input/output address bus*. The CPU also performs all of the arithmetic calculations and memory storage/retrieval operations. The *memory address bus* defines a specific area in the memory to be worked upon, and the *memory data bus* either stores or retrieves data from that specific location.

210

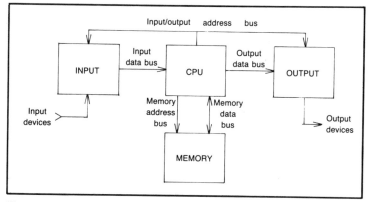

Fig. 18-1. Basic computer block diagram.

The *output section* accepts the digital information from the CPU, converts the information into a usable form, and routes it to the appropriate output device.

As stated previously, a computer consists of four basic parts: the input, output, CPU, and memory. The central processing unit (CPU) can be further divided into the *arithmetic logic unit* (*ALU*) and the *read only memory* (*ROM*). The ALU controls the logical steps and order for performing arithmetic functions. It will interact with the ROM for instructions for performing redundant operations. The ROM will also contain instructions pertaining to start-up and power loss conditions, and instructions for conversion of higher-level languages to *machine-language programs* (*MLP*). The CPU usually contains the *real time clock* (*RTC*). The RTC is used to cycle the CPU, and time the real-time programs as written by the user.

The success of the computer is governed by the speed of its operations, not its complexity. Basically, a computer is only capable of adding, subtracting, and accumulating data. Because it is capable of performing these simple operations at amazingly high speeds, complex mathematical calculations can be broken down into simple steps which the computer can then calculate. For example, a computer actually multiplies by redundant addition of the same number. Division is accomplished by redundant subtraction.

The smallest single operation performed by a computer is the *machine cycle*. This consists of two stages: the *fetch cycle* and the *execute cycle*. During the fetch cycle, the processor fetches an instruction from memory. Then, during the execute cycle, the computer performs some action based upon the content of that instruction. The processor knows which instruction to go to next from the address stored in the program counter. It always contains the address of the next instruction. When program instructions are written, they are arranged in a sequential order, and the program counter simply increments by one for each machine cycle.

If the central processing unit is contained in a single integrated circuit, the IC is referred to as a *microprocessor*. A computer based on a microprocessor chip is called a *microcomputer*.

Modern process controllers, utilizing microprocessor-based control, usually consist of a *single-board computer*. A single-board computer is a microcomputer with its input, output, CPU, and memory all physically located on one printed-circuit-board assembly.

PROGRAMS

Basically, the components in a control system can be divided into two categories: *hardware* and *software*. Hardware is all of the physical components required to make up the system. Software is the term given to the programs written for a computer system.

There are many different program languages in common use today. The reason many languages exist is due to the need for *procedure-oriented languages*, which allow the programmer to concentrate on the problem to be solved rather than the difficulties of writing the program. Procedure-oriented languages are often referred to as *high-level languages*. In general, the easier a program is to understand from a human point of view, the higher its level.

For smaller computers and microcomputers, the most common programming language is referred to as BASIC. BASIC language will vary from one computer manufacturer to another, but the general concept remains the same. BASIC consists of symbol words (abbreviations) which closely resemble the operation desired by the programmer. *Machine language* (MLP) is the most difficult language to use. It consists of the actual binary words the computer needs to perform the desired operation.

Programmable logic controllers (*PLC*) are specifically designed for use in industrial control. They utilize a special type of BASIC oriented toward the input/output functions of the PLC. This language is closely tied to the nomenclature used in industry to describe relay ladder logic diagrams.

Since the primary use of PLCs is fairly well defined, the internal ROM programs can be extended to a much higher degree than in a general-purpose computer and, therefore, simplify the user program to a great extent. If the user is already familiar with relay ladder logic, he will be able to program a PLC with very little training. The ROM program may also contain specific programming to generate documentation in common relay ladder logic formats.

A common high-level language used in computerized numeric control (CNC) systems is called *APT* (*Automatically Programmed Tools*). This language contains hundreds of instructions used for control positioning, continuous-path, and contour movements of a machine tool. These instructions can be classified into four categories: *auxiliary, postprocessor, geometry,* and *motion.*

Geometry instructions—Use a coordinate system to define the locations of surfaces and points on a workpiece.

Motion instructions—Describe the path of movement of the machine tool.

Postprocessor instructions—Define parameters of the machine tool, such as speed and feed rate.

Auxiliary instructions—Miscellaneous optional commands.

Other common high-level programs are FORTRAN, PASCAL, and COBOL.

COMPUTER CONTROL OF ANALOG SIGNALS

When a computer needs to examine an analog signal, the analog signal must be converted to digital words before the computer can understand it. A logic system designed for this purpose is called an *analog-to-digital converter* (commonly symbolized by A/D). The basic concept of A/D conversion is based on a principle called Nyquist's Sampling Theorem, which states:

If an analog signal is uniformly sampled at a rate at least twice its highest frequency content, then the original signal can be reconstructed from the samples.

In other words, if you take numerous instantaneous readings of an analog signal and store these readings (in the form of digital words), you may reconstruct a close representation of the original analog signal by placing the instantaneous readings in their original sequential relationship.

An A/D converter performs the first half of this process by converting the analog signal to numerous words at some preset periodic rate. The digital words can then be stored or manipulated by the computer.

To reconstruct the original analog signal, or any analog signal, a *digital-to-analog converter* (symbolized by D/A) is required. The D/A converts digital words from the computer to analog voltages and outputs them at the rate at which they were originally collected by the A/D. In this manner, a computer may read or output any analog signal. Most importantly, it can also analyze and manipulate in the same way. This is the basis for all microprocessor-based process controllers.

TROUBLESHOOTING COMPUTER SYSTEMS

1. Virtually all computerized systems have some form of self-diagnostic capability. In other words, the computer can help isolate the problem by means of its own processing abilities. In some systems, diagnostic programs reside permanently in the computer's ROM memory. In other systems, special diagnostic programs must be loaded into the system whenever a system malfunction occurs.

2. If the system is totally inoperable, you should suspect a power supply problem. As with analog ICs, digital ICs must have a power source to operate.

   Verify that correct voltages are present. If the proper power-supply voltages are not present, there are basically two possibilities: either the power supply itself is defective, or one of the devices or printed circuit boards it is powering has an internal short. This can cause the power supply to look defective.

   Virtually all digital systems incorporate regulated power supplies with internal current limiting. If the current drawn from the power supply exceeds this specified current limit, the power supply shuts down to prevent further damage.

   The easiest way to isolate the problem to either the power supply or an external system component is to disconnect the power supply outputs. If the power supply voltage(s) return to normal, begin reconnecting one system component at a time until the power-supply problem returns. In this manner, the defective system component can be isolated.

   If the system utilizes plug-in printed circuit boards, pull out one board at a time to see if the power supply returns to normal after the removal of a specific board. If so, that board is defective. Remember, to turn off the system power before removing or inserting any system component or printed circuit board.

3. Don't forget to verify that all external inputs and outputs to the system are functioning properly. These inputs may include switches, proximity sensors, photoelectric eyes, or auxiliary contacts.

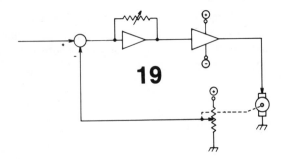

19

Industrial Control Fundamentals

THERE ARE BASICALLY THREE TYPES OF INDUSTRIAL PROCES-
ses that encompass all industrial manufacturing systems. These
include:

1. *Continuous processes*. Raw materials enter one end of the
   system, and the finished product emerges from the other
   end. During this time, the process runs continuously.
2. *Batch processes*. A set amount of each of the inputs to the
   process is received in a batch, and then some type of operation is performed on the batch to produce a finished
   product.
3. *Discrete processes*. Individually handled and manufactured
   parts, which do not have a set rate at which to move
   through the process, comprise this system.

INDUSTRIAL CONTROL SYSTEM BASICS

The major components of any industrial control system are sensors, signal conditioners, controllers, and actuators.

Sensors

Sensors normally provide the inputs the process uses to control the manufacture of the product. Sensors can be divided into
two categories: process sensors and external sensors. *External sen-*

sors are the inputs from the human operator that define the manufacturing parameters. *Process sensors* are inputs that tell the controller whether or not the manufacturing parameters are being met. Sensors may sometimes be referred to as *transducers*.

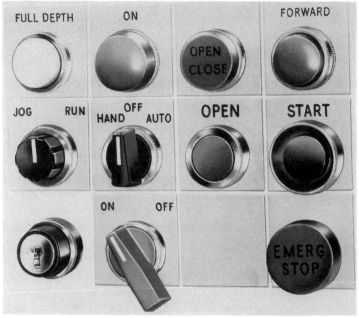

A typical industrial control panel functioning as the operator to machine interface. It includes operator controls along with indicators to inform the operator of how his process is running (courtesy of Micro Switch, a Honeywell Division).

Temperature sensors commonly used for process control applications include thermocouples, thermistors, RTDs, and IR thermometers. A *thermocouple* consists of two junctions of two dissimilar metals which generate a small voltage proportional to its temperature. The output of a thermocouple is linear. The thermocouple is the most common temperature sensor used in industry.

The *thermistor* is made of semiconductor material, and its operation is based on the negative temperature coefficient inherent to all semiconductor materials. A thermistor's resistance is inversely proportional to its temperature. This resistance is converted to a voltage for input to the signal conditioner. A thermistor's output voltage is nonlinear. The thermistor is used in applications requiring high sensitivity.

The *RTD* (*resistive thermal device*) has a positive temperature coefficient. It also has high linearity, and high stability over long periods of time.

IR (*infrared*) *thermometers* measure the infrared radiation emitted from all materials. The amount of radiation is directly proportional to the material temperature. IR thermometers are useful in applications in which it is impractical to physically contact the substance to be measured.

Pressure sensors include capacitive, barometric, and strain gauge types. A *capacitive pressure sensor* consists of an evacuated chamber with two ends (plates) which are insulated from one another. As pressure changes occur, these two ends move closer together or farther apart, causing a change in capacitance. This change in capacitance can be measured in various ways to provide a signal proportional to pressure.

A barometric pressure sensor also consists of an evacuated chamber and a plate (or diaphragm) that physically moves in or out with a change in pressure. This in-and-out movement of the diaphragm is mechanically coupled to an electronic device (usually a potentiometer) to convert the physical action to an electrical signal.

Strain-gauge pressure sensors are made from a piezoresistive semiconductor material, which changes its resistance when distorted. The piezoresistor is attached to a weigh table or evacuated chamber in such a way to produce stress or distortion when a pressure is applied.

Two common instruments for measuring flow are the turbine and the differential pressure cell. A *turbine* is a propeller placed in line with the flow, causing it to rotate at a speed proportional to the flow. This rotation can be sensed by driving a small generator with the propeller (or impeller) or proximity sensing through the wall of a pipe. Operation of a *differential pressure cell* is based on the Venturi principle, which defines the relationship of pressure, velocity, and kinetic energy within a closed-flow system. By monitoring the pressure differential at a pipe restriction, a corresponding flow measurement can be calculated.

Photoelectric sensors, limit switches, capacitive switches, and Hall-effect switches are all examples of position sensors. A *photoelectric switch* consists of a light emitter and a receiver. When an object breaks the beam of light, an output is given. *Limit switches* are simply electromechanical switches activated when an object comes in physical contact with them. *Capacitive switches* are used to sense the presence of metallic objects by allowing an object to be one plate of the capacitor. *Hall-effect switches* utilize the Hall-effect principle to sense the presence of metallic objects. The Hall-effect principle is a relationship of current and voltage when in the presence of a magnetic field.

A few examples of typical industrial switches (courtesy of Micro Switch, a Honeywell Division).

Industrial proximity sensors used for non-contacting sensing applications (courtesy of Micro Switch, a Honeywell Division).

Signal Conditioners

Input signal conditioning is necessary to convert electrical signals from sensors to electrical signals that the controller can understand or process. *Output signal conditioners* convert outputs from the controller to voltages or currents compatible with the output actuators.

Controllers

The *controller* is the brain of the system. It receives the inputs from the sensors and performs mathematical calculations and/or logical comparisons to decide what must be done next. It then generates the correct output signals to carry out the decision. Controllers can be divided into two basic categories: sequential controllers and process controllers. *Sequential controllers* are used in manufacturing processes requiring only ON/OFF control signals. These processes may be very basic but may also include complex sequential control, as used in NC, CNC, and multifaceted handling operations. Sequential control also overlaps into the programmable logic controller (PLC) field. *Process controllers* are used to control continuously variable processes, such as sewage treatment and chemical manufacturing processes.

The most basic form of sequence controller is the *electromechanical drum timer*. The *drum* is a cylinder rotated by a *timer* motor. The drum contains projections or indentations to actuate switches as the timer motor rotates the drum. The projections or indentations can be adjusted for the particular process. One complete rotation of the drum is, therefore, one process cycle. Solid-state versions of the drum timer include discrete sequential timers and sequential type programmable controllers. Sequential controllers are almost always incorporated into open-loop control systems.

Actuators

Actuators are devices that convert an electrical output control signal to a physical action. Commonly used actuators are the solenoid, relay, and motor. A *solenoid* is a device that produces a straight-line mechanical force when electrical power is applied to its associated coil. A *relay* consists of a solenoid mechanically connected to an electrical switch. *Motors* convert electrical power to rotational force.

CONTROL SIGNALS

The types of signals used in control systems can be classified into one of two types: analog or digital. *Analog control signals* are continuously variable, and either the signal amplitude or the frequency is directly related to a sensor measurement or actuator position. *Digital control signals* can range from a simple on/off relay contact action to complex digital words containing as many as 16 bits.

NOTE: A *bit* is one binary digit (either a 1 or 0). For example, a digital word such as 1010 contains four bits. When multiple bits are used together to form a digital word (usually 8 or 16 bits long), the digital word is referred to as a *byte*.

A digital process reporter used primarily in the pharmaceutical industry (courtesy of Honeywell Process Control Division, Fort Washington, PA 19034).

OPEN-LOOP AND CLOSED-LOOP CONTROL SYSTEMS

Control systems can be structured as either open-loop or closed loop. An *open-loop control system* calculates the result of the output. It is free of feedback to ensure that the desired results are being obtained. An example of an open-loop system is shown in Fig. 19-1. The external inputs from the operator specify the actual desired result from the process.

These external inputs are signal-conditioned and sent to the controller. The controller performs the required calculations and sends its output to the output-signal conditioners, which provide the appropriate control signal to the process actuators. The process actuators provide the physical control of the process.

A good example of an open-loop control system is a microwave oven. The external inputs from the operator are the desired time and heat settings. The pushbuttons used by the operator are signal-conditioned to the appropriate voltage signals needed by the controller. Depending on the particular type of microwave oven being

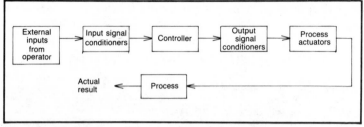

Fig. 19-1. Block diagram of an open-loop control system.

used, the controller may be as simple as a few relays and a timer. The output of the controller is signal-conditioned to provide the correct power for the desired heating in the correct time duration. The process actuator is the magnatron tube providing the microwave radiation. The process is the cooking of the food placed in the oven. If the operator enters an incorrect time or heat setting, the food will not be cooked properly.

In a *closed-loop system*, process sensors are used to provide a feedback signal to the controller. This allows the controller to monitor the process output and make any changes necessary to achieve the desired actual result. A block diagram of a closed-loop system is shown in Fig. 19-2.

Refer to the previous example of the microwave oven. If you were to insert a temperature probe into the food, the temperature probe would provide a feedback signal to the microwave oven controller to automatically turn itself off when the food was properly cooked. This is an example of a closed-loop system, because the

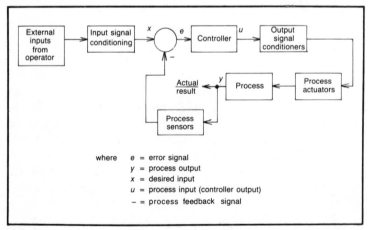

Fig. 19-2. Block diagram of a closed-loop control system.

cooking process is actually being monitored by the temperature probe. The temperature probe would be the process sensor.

CLOSED-LOOP CONTROL SYSTEMS

The previous examples describing the action of a microwave oven are good for developing the basic concept of the difference between open-loop and closed-loop control, but they are not representative of a true industrial continuous process. The factors involved in controlling a continuously varying process become much more complicated.

Figure 19-3 shows the basic components involved in a simple temperature control system. Assume the desired input (X) is 5 volts. For the sake of discussion, you can consider this 5-volt desired input signal to be representative of attempts to maintain the temperature of the chemical in a heater tank at precisely 200 degrees. The process sensor is a temperature sensor with an analog output voltage relative to the chemical temperature. If the actual chemical temperature is 200 degrees, the process feedback signal (-) would also be 5 volts. Therefore, the error signal (e) would be zero, and the output to the heater would not change.

Assume that the actual chemical temperature dropped to 160 degrees. The process feedback signal will drop to 4 volts, causing the error signal to increase. This causes more heat to be applied to the tank to try to bring the chemical temperature back up to the desired 200 degrees. As the chemical temperature increases closer to the desired 200 degree point, the error signal will decrease

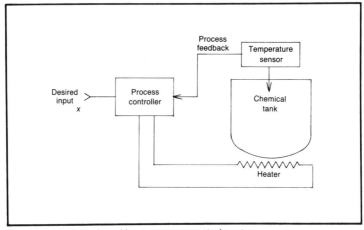

Fig. 19-3. Typical closed-loop process control system.

proportionally. Thus, this type of control action is referred to as *proportional*.

The amount of change in the error signal relative to the difference in the desired input signal and the process feedback signal is called the *system gain* (symbolized as the K factor). The system gain is often referred to as the *proportional band*. As the system gain is increased, the proportional band decreases.

Proportional control has a serious drawback. In the previous chemical heater example, the actual chemical temperature would never reach the desired 200 degree point. Since the error signal decreases as the chemical temperature gets closer to the desired 200 degree point, the error signal would eventually become so small that the process controller could not act upon it. If you tried to increase the system gain (K), the process would become unstable and would oscillate.

For this reason, proportional control is normally used in conjunction with another type of control action called *integral control* (sometimes called *reset action*). Integral control continually adds the error signal relative to time. For example, if an error signal happened to be .1 volt and the integral control was adjusted to 1 second, after 5 seconds the error signal would be .5 volt. The integral control will continue to add the error signal until the continuous offset from proportional control alone is removed.

A third type of control action is called derivative. *Derivative control action* acts upon the rate of change in the process, rather than the difference between the desired input and process feedback. For example, if the process changes slowly, the corrections will be made solely by the proportional control action. On the other hand, if a very rapid process change occurs, it can be assumed the rapid change will continue until the process is out of specification. In this case, the derivative action would actually tend to overcompensate with a dramatic, nonproportional change in the error signal in an effort to compensate for the eventual anticipated error. For this reason, derivative action is sometimes called *anticipatory control*.

Most modern process controllers incorporate all three control actions and are, therefore, called PID (proportional-integral-derivative) controllers. PID controllers have separate controls to adjust for each control action desired. Many modern, programmable logic controllers also contain internal PID loops.

A PID temperature/process controller with a 1″ high digital readout (courtesy of Love Controls Corp., 1475 S. Wheeling Rd., Wheeling, IL 60090).

This PID temperature controller is microprocessor based, contains its own self-diagnostic routine, and is available with plug-in options (courtesy of Honeywell Process Control Division, Fort Washington, PA 19034).

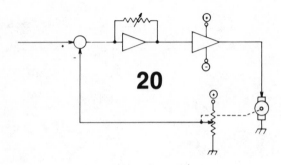

20

Continuous Process Control

THIS CHAPTER WILL CONCENTRATE ON MODERN PROCESS-CONtrol methods and terminology. In virtually every large, continuous, process-control application, a computer is the process controller of the system. The computer may be a *programmable-logic controller* (*PLC*) or some other specially adapted industrial computer. Programmable logic controllers will be discussed separately in a later chapter.

Before studying an example of a typical continuous process-control system, you should become familiar with the following terms and definitions.

DEFINITIONS

Process variable—A process parameter that can change value.

Controlled variable—A process variable controlled by the process control loop.

Manipulated variable—The variable changed by the controller to maintain the correct value of the controlled variable,

Disturbance—The parameters that affect the controlled variable but are not controlled by the process controller.

Feedback gain—The proportional band.

Amplitude proportional gain— A modifier to the feedback gain which increases linearly as the process error increases.

Total gain—The product of the feedback gain and the amplitude proportional gain.

226

Feed-forward gain—Target changes.

Load variable—A disturbance caused by a change in demand of the controlled variable.

Setpoint—The desired value of a controlled variable.

Excessive error—Limits that guard against feedback signal failure or excessive process changes.

Dead band—The acceptable control error allowed before corrections are made.

System lag—The total amount of time delay in a control system from the time a manipulated variable is changed until the controlled variable responds.

Actuator control—A type of control that minimizes physical and mechanical peculiarities in each actuator of a control system.

Process partials—The calculated interaction between numerous controlled variables in large process-control systems.

Setpoint limits—Specified limits in the process controller that are compared with actuator setpoints for the purpose of monitoring actuator failures.

Alarm message hysteresis—A type of filter (i.e., time delay) that prevents repeated triggering of alarm messages due to system noise.

Alarms—External actuators to notify the operator of control or system problems.

Alarm band—The percent of deviation allowed before the operator is notified of a problem.

Decoupling—Mathematical calculations for actuator changes considering the interactive nature of several controlled variables.

Backlash—A mechanical characteristic of an actuator that will cause a response change if the actuator is required to reverse direction.

Slewing rate—The amount of actuator change per output time.

Dead time—Actuator hysteresis caused by pressure buildup of mechanical linkage.

Maximum on-time/off-time—The minimum and maximum time periods in which the process controller is allowed to change an actuator position in one control move.

Modeling—A software simulation of a machine response.

Cascade control—A control scheme in which two or more complete control systems interact for total product manufacture control.

Time constant—The time period from the start of a change until the change reaches 63 percent of its final value.

Transport lag—The same as transfer lag. It is made up of the dead-time delay and the transport delay.

Transport delay—The speed-dependent term of the transport lag.

To relate the preceding terms to an actual process, consider a hypothetical plastic-extrusion process while the terms are described and analyzed. This plastic extrusion process is shown in Fig. 20-1.

You must first melt plastic pellets and then force them through an extruder head under high pressure. The feed rate of the molten plastic is controlled by the speed of a rotating, threaded rod, called the screw, located in front of the extruder head. From the extruder head, the hot plastic passes through heated chrome rollers that form it into a flat, symmetrical sheet. This sheet is then measured with a caliper sensor to ensure that the proper thickness is obtained. See Fig. 20-1. From there, the sheet continues down a conveyor, where it is eventually coated with a 1-mill layer of special high-tensile strength Mylar. After the coating process, the sheet is cooled, and the finished product is again measured for proper thickness.

This complete process can be divided into two sections, or phases: the extrusion phase and the coating phase. In the *extrusion phase*, the process variable is any variable capable of changing. The controlled variable is the thickness (caliper) of the plastic sheet. The manipulated variables are the stock (pellets) feed rate (controlled by the screw speed), the pressure of the extruder head, and the speed of the chrome rolls.

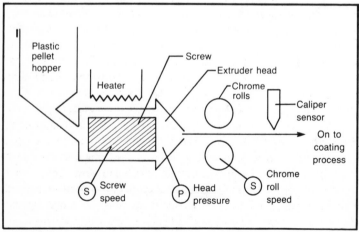

Fig. 20-1. Basic plastic extrusion process.

A disturbance could be caused by defective pellets coming from the pellet-feed hopper. If a large disturbance occurs, the *amplitude-proportional* gain will increase the *total gain* greatly to improve the speed at which the process returns to the setpoint. The total gain will decrease to the *feedback gain* when the process returns close to setpoint value.

The speed of the process is the load variable. Assume you have adjusted the *setpoint* for a 40-mill plastic sheet. If the plastic varies plus (+) or minus (−) .25 mill, no correction will be made by the *process controller*. This is the *dead band*.

Assume you change the setpoint to 42 mills. You observe it takes about 30 seconds for the actual caliper to begin increasing. This is the *transport lag*. Of this 30-second delay, it takes 5 seconds for the screw speed to begin increasing after a control output is applied to it. This is the *dead-time delay* of the screw speed control. Once it starts to increase, it increases at a rate of 5 percent per second, which is its *slewing rate*.

Now that the screw speed has increased the feed rate, it takes 10 seconds for the increase of molten plastic to appear at the chrome rolls. In other words, it took about 15 seconds for the controlled variable to respond. This is the *system lag*. It will take another 15 seconds for the thicker plastic to actually appear at the caliper sensor due to the speed at which the machine is running. This is the *transport delay*. The *process time constant* is the time period from the beginning of the change to the point where the change is 63 percent of what it eventually will be.

The actuators in the extrusion phase of the process are the screw speed control (pellet-feed rate), the extruder head heaters, the extruder pressure valve, and the motor speed control for the chrome rollers. Each of these actuators is controlled by its own independent control loop, called the *actuator control*. The process controller sends out targets (or setpoints) to these control loops. The factor by which these targets are multiplied when they are sent to these individual control loops is called the *feedforward gain*.

The individual loop controllers are monitored by the process controller. The process controller will not allow massive changes to be made in the process in only one control move, so it defines maximum on-times to limit the effect of any individual loop. If any loop actuator moves an excessive amount or shows an excessive error, a setpoint limit may be exceeded. If this error is outside the alarm band, the operator may be warned by an alarm message.

You decide to make another setpoint change to 35 mills for a new product run. The control loop for the screw speed must now reduce the pellet-feed rate. To do this, it must know the direction of the last control move to compensate for backlash of the loop. (This backlash is caused by mechanical slop in the motorized potentiometer used to provide a speed-control signal to the screw-speed motor control.) Since it must reverse direction from the last control move, it will compensate with a longer on-time.

Written into the process controller's software is a model of this machine. Therefore, the process controller knows how the machine will respond to the last setpoint change. It knows that there is an interactive relationship between chrome roll speed, feed rate (screw speed), and head pressure.

As the process controller reduces the thickness down to 35 mills, the actuator decoupling provides compensation to the individual interactive control loops to minimize process disturbances. This complicated procedure is necessary in the event that several manipulated variables must be changed at one time, which they usually are.

The extruder process controller now sends out a new target for the coater process controller to compensate for the change in caliper. Since the coating process is an independent control system by itself, the interaction of the extruder control and the coater control is referred to as *cascade control*.

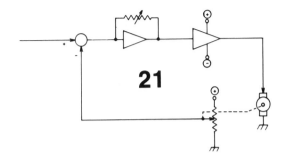

21

Jobshop System Control

JOBSHOP SYSTEMS ARE AUTOMATED SYSTEMS THAT CONTROL a sequence of operations performed on the feedstock (raw material). You may also refer to jobshop control as semicontinuous manufacture. The common types of jobshop control systems are referred to as *numerical control* (NC), *computerized numerical control* (CNC), and *direct numerical control* (DNC).

Basically, an NC system is a manufacturing system that uses a program of instructions to control a sequential set of operations performed by a machine. A typical NC system can be divided into three parts: the program, the machine controller, and the machine. The more common types of machines used in an NC system are lathes, milling machines, punch/presses, welders, and drills. (In many cases, the machine is called the machine tool.)

The earliest forms of NC control used a paper tape to hold the program. The paper tape contained a series of punched holes that described each instruction. At the beginning of a manufacture cycle, the paper tape was loaded into a paper tape reader to convert the hole positions into a type of digital word describing the desired operation. These operations would be performed one at a time in sequence. When each operation was completed, an acknowledgement was sent back to the paper tape reader to advance the paper tape to the next instruction.

It should be obvious at this point that all operations carried out by the NC system had to be sequential, because there was no way to randomly access the paper tape. It had to be read one instruction at a time and could not advance until that operation had been physically carried out. Also, the entire paper tape had to cycle once for each time the program was performed.

The controllers used in this generation of equipment were limited in flexibility. They could only work with one kind of language and one type of machine. Although some of these systems could recognize the successful completion of certain operations, the majority simply recognized the completion of the instruction. Also, the operator-to-machine interface was limited in its ability to communicate. For this reason, the operators had to be well trained on the machining operations in the event the system had to be put in manual mode to manually correct for inconsistencies in the manufacturing cycle.

More modern NC systems are widely used today. They maintain a rigid sequential format just like their ancestors, but the controller is an intelligent processor. In other words, it has memory capability. Since the program can be stored in the processor's own internal memory, the ease of writing the programs has increased, because the programmer does not have to maintain the proper timing and sequence required in the older type NC systems. Many systems still use paper tape as the hard copy and the means of loading the program into memory, but the information contained on the tape is not necessarily in sequential order. Also, the paper tape is not required to cycle while the NC system is running.

Because the ease of programming these second-generation NC systems has improved, it is cost-effective to write more sophisticated programs to monitor, in more detail, the actual parts manufacture. The operator-to-machine interface was also improved upon. Due to these improvements, the qualifications of the operators were not as critical and the simplicity of operation was improved upon.

However, these second-generation NC systems have some disadvantages. They are very expensive. Also, hardware complexity has increased, causing service and reliability problems. Finally, they are not much more flexible than the older systems.

In the latest NC systems, the controller has been replaced by a true computer. These systems are referred to as computerized numerical control (CNC). Because of the recent advancements in integrated-circuit technology, general-purpose computers can be mass-produced and sold very inexpensively. These general-purpose computers, due to their flexibility, can be used in a variety of CNC applications. In other words, a milling machine, a drill press, a four-axis lathe, and a 500-ton punch may all use the same general-purpose computer as their controller. This general purpose philosophy has done much to improve hardware costs, as well as reliability and servicing.

Nonvolatile (hard-copy) program storage materials used in CNC systems include paper tape, magnetic tape, and diskettes. The programs themselves are much more complex than the second-generation NC systems. The high-level mathematical manipulation capabilities of the computer are used for more complex machine-contouring functions. High-level programming languages (such as APT) can be utilized. Because of the high-level operator-to-machine communications, operation can be extremely simple. Also, the modern system diagnostic capabilities allow the computer to diagnose internal system problems.

The extensive capabilities of the computer have allowed state-of-the-art CNC systems to incorporate *adaptive control*. Adaptive control gives the CNC system the ability to adapt its operation if certain disturbances occur during manufacturing. For example, if the cutting tool should become dull on a CNC lathe, the adaptive control may reduce the spindle speed to keep from overheating the cutting tool.

The evolution of the CNC system has brought about the direct numerical control (DNC) system. In a DNC system, one large central computer can control as many as 256 CNC machines. The DNC computer supervises and controls the real-time operation of the multiple CNC system and is useful for data reporting and analysis purposes.

OTHER TYPES OF JOBSHOP CONTROL

There are methods other than CNC and NC systems to manufacture (or handle) discrete parts on a high-volume basis. The reasons for the existence of these other methods are as follows:

1. The application may be unique. Specialized systems compatible with the application may not be available.
2. It may be cost-effective to upgrade older existing equipment.

As in the case of NC and CNC, you may divide these systems into three main sections: the program, controller, and machine.

Two basic types of microprocessor control methods are used

An example of a general purpose single board computer (courtesy of Wintek Corporation, Lafayette, Indiana).

for upgrading systems and for installing specialized manufacturing systems. They are the *single-board computer* and *programmable logic controller* (PLC).

The PLC provides the systems-design engineer a relatively inexpensive general-purpose controller. The I/O system of a PLC is usually optically isolated and used for general purposes. It accommodates virtually any type of input or output device. The program consists primarily of relay-ladder logic or simplified data logic requiring very little programmer training. In other words, the PLC is the quick and easy answer to most custom automation requirements.

The single-board computer is gaining in popularity due to its general-purpose nature. The single-board computer requires much more system-development time because the design engineer must design the I/O hardware, develop the software (usually in basic language), and provide the power supply. The advantage to the single-board computer is its cost (as low as $100.00) For a single application, the end cost of a single-board computer system would be much higher than a PLC system because of the increased systems-development time. But for multiple systems, the cost savings could be significant.

JOBSHOP MACHINE INTERFACING

There are many methods of interfacing a modern controller to a machine tool. Following are the most commonly used methods.

The 4-20 Milliamp Current Loop. The EIA and ISA standard for a current loop range is 4 to 20 milliamps. The 4 milliamps represent the zero point, and the 20 milliamps represent the full-scale point. A current loop is used to eliminate the line losses associated with the use of voltage or resistance as the analog output from a process sensor.

When interfacing a current loop, the current is usually converted to a voltage to be read by the interface. This can be accomplished easily by using a precision resistor or potentiometer. A 250-ohm resistor will convert 4 to 20 mA to 1 to 5 volts. Other resistor values will convert the current to a proportional voltage so that virtually any interface range can be accommodated.

The Speed/Ratio Method. Speed and ratio (the comparison of two or more speeds) can be measured by counting pulses from an incremental encoder for a defined period of time (usually 1 second). An incremental encoder outputs a specific number of pulses for each rotation of its shaft. The number of pulses per time period will define the relative speed at which the encoder shaft turns.

If two encoders are used, the two speeds can be compared to create a ratio. Most PLC manufacturers offer an optional high-speed

A programmable logic controller incorporating a unique register access module which allows the operator to change register values without process shutdown (courtesy of Eaton Corporation, Cutler-Hammer Products).

count module which will directly interface to many incremental encoders. The various encoder manufacturers also provide *frequency-to-voltage converters* so that an analog voltage may be interfaced instead of a frequency.

Tachometer generators may also provide the analog voltage for measuring speed or ratio. A tachometer generator is simply a small

A versatile temperature and current loop calibrator/monitor (courtesy of Biddle Instruments, Blue Bell, PA 19422).

ac or dc generator. Its analog output voltage is proportional to the rotational speed of its shaft.

The Temperature Method. Temperature is usually measured with thermocouples or RTDs. Each of these devices requires a *signal conditioner* before it can be interfaced. Signal conditioners convert the extremely small voltage and resistance changes to high-level voltages usable by the controller I/O.

The Analog I/O Method. Virtually any analog input or output can be interfaced as long as it is at some reasonably high level (1 volt or higher). Voltages below 1 volt are usually amplified before they are applied to the controller I/O. Voltages above 10 volts can be easily attenuated by a simple voltage-divider network.

The Discrete I/O Method. The *discrete (or digital) I/O* is used to sense or output on-off functions. These include home position, position complete, strobes, data words from absolute encoders, load, store, increment, etc. The discrete I/O of a PLC system refers to the on-off functions analogous to a relay-logic system.

BASIC MACHINE INTERFACE PHILOSOPHY

The perfect machine interface would not be speed-dependent, nor would it incorporate any physically contacting sensors. There would be no moving parts, and nothing could be damaged by adverse environmental conditions. Of course, this is not realistic, but modern control components come reasonably close to this ideal state.

Optical isolators provide complete isolation between the computer and the real world. Optical isolators are solid-state devices containing no moving parts or contacts and are commonly used to isolate the computer I/O from machine sensors.

Modern *infrared modulated photoelectric sensors* are not affected by ambient light, contamination, extremes of temperature, or mechanical shock. They are noncontacting and, for the most part, do not contain any moving parts (although some photoelectric sensors may contain an output control relay). ''Light-operate'' and ''dark-operate'' modes simulate normally open and normally closed contact conditions.

Transformers are used to isolate ac inputs. They can also isolate some types of analog inputs (frequency and pulsed).

Solid-state relays are actually optically isolated triac or SCR control devices. Their inputs accept a wide voltage range (typically 3-30 Vdc). They can handle high-power applications and have no moving parts or contacts to wear out. Solid-state relays are often used for discrete interfacing.

Inductive proximity switches are a noncontacting method of sensing the presence or position of ferrous metal objects. Most of these units operate on an inductive reactance principle and are solid-state devices with no moving parts. For applications requiring the sensing of ferrous metal objects (machine cams, gear teeth, etc.), inductive proximity switches are preferred to photoelectronic eyes because they are not adversely affected by dirt and contamination. Proximity sensors are also available for the detection of nonferrous metals. Capacitive proximity switches are also available for detection of the presence of virtually any material.

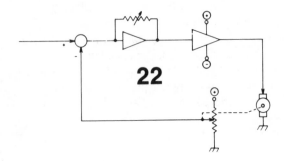

Programmable Logic Controllers

PROGRAMMABLE LOGIC CONTROLLER (PLC) IS THE CURRENTLY used term for the programmable controller. The term *programmable controller* (PC) was dropped because of the confusion of this term with the term "IBM PC," in which PC stands for *personal computer*.

A PLC is a computer specially designed for one-of-a-kind industrial applications or equipment-upgrading projects. Since most industrial design personnel are already familiar with relay-ladder logic, the PLC uses this same program language to make it easy for industrial personnel to use. In addition to standard control-relay functions, the PLC is capable of creating timers, counters, and drum controllers. For more complex requirements, some PLCs offer an optional data-handling capability for interfacing high-speed counters, encoders, computers, peripherals, and analog devices. For process-control applications, some PLCs include PID control routines in the internal ROM program.

The I/O (input/output) structure of most PLCs is *user-definable*. This means the user can define the inputs and outputs to suit the application. Once the I/O has been defined, the user simply plugs in the appropriate I/O modules. The I/O modules contain the necessary interface to the processor, together with protective optical isolation. The inputs and outputs from the actual process connect to these I/O modules, so they must be appropriately rated for the desired voltage or current required. Each I/O module may contain as many as eight inputs or outputs. In most cases, inputs and out-puts will not be mixed on the same I/O module.

Because power failures and brown-outs (in which the ac power drops to an abnormally low value) are common within industrial manufacturing facilities, PLCs contain some form of nonvolatile memory system. Older PLC systems may contain battery backup to supply power to a volatile memory system. Most newer systems contain either PROM (programmable read-only memory) or EAROM (electrically alterable read-only memory) for this purpose.

Additional hardware that may be purchased with a typical PLC system includes program loaders, remote operation panels, printer interfaces, and special function I/O modules (such as stepper mo-tor drives, analog setpoint inputs, multiplexed analog inputs, etc.). System interfaces, which allow the interconnection of several PLC systems, are also available along with I/O simulators for program development.

DEVELOPING THE PROGRAM

All PLC systems include some form of external programming device called a *programmer*. This programmer may be a small hand-held unit (about the size of a calculator) for smaller systems or a complex, microprocessor-controlled video unit for larger systems. In either case, the programmer is not intended to be permanently attached to the processor unit. Once the program is developed and debugged, the programmer is removed and stored for future use. If the programmer has a monitor capability (the ability to watch the program in actual operation), it may be used as a valuable troubleshooting aid.

The first step in preparation for writing the program is to make an I/O listing containing all of the inputs to the PLC and all of the outputs coming from the PLC. (This I/O listing should be saved as a valuable troubleshooting aid.) Once this has been completed, the inputs and outputs must be defined as to their physical con-nection to the I/O. When this has been accomplished, you should verify that the correct I/O modules are being used with the appropri-ate inputs and outputs.

After you have made an I/O listing and defined the I/O, you can begin to write the program. The program is written in relay-ladder logic as if actual relays were to be used. Naturally, you must know the logical functions you want performed. Inputs to the PLC will be shown as *relay contacts*; outputs from the PLC will be shown as *relay coils*. These inputs and outputs are labeled according to the I/O definition.

The majority of the relay-ladder logic will be performed by the *internal relays*. These internal relays are actually data bits being controlled by the processor logic and the user program. You consider them as being relays for the ease of writing the relay ladder-logic program. A typical PLC may have several thousand internal relays. Internal timers and counters are also available for the user program.

DATA PROGRAMMING

A relay-ladder program is often referred to as a *discrete program* because it is written as if discrete relays were being used. In more complex programs, you often need to utilize a higher level of mathematical functions called the *data programming ability*. Larger PLCs have this capability.

A PLC with data programming ability is actually a hybrid controller. In other words, it is a PLC with some standard computer operations. For example, without data programming ability, a PLC cannot read a 16-bit parallel word. Common mathematical functions included in a typical PLC with data programmability are addition, subtraction, division, multiplication, comparison, and shift-register actions.

Data programmability enables you to read data and use this data as a complete word. This word may represent a count, absolute encoder position, external keyboard entry, analog voltage, etc. It is stored in a register (8-bit or 16-bit) which may then be acted upon by any mathematical function or discrete logic function. Results of mathematical operations may be stored in other data registers.

Most PLCs use special data input and data output modules for the data I/O. Some PLCs have separate discrete and data I/O ports.

OTHER SYSTEM FUNCTIONS

Larger PLC systems have the capability to communicate with computers and printers. This is usually accomplished with a standard serial communications method (either RS-232 or RS-422). Some systems can even be programmed to display video messages. Multiple PLC systems can be interconnected, and the entire system can be controlled by a supervisory computer.

All PLC manufacturers include detailed user manuals with their

This PLC can support analog I/O in conjunction with discrete I/O. It is also capable of data handling and double precision math functions (courtesy of Eaton Corporation, Cutler-Hammer Products).

PLC systems. These manuals are generally very easy to use and understand even for those who have never used a programmable logic controller.

DISCRETE PLC INPUTS AND OUTPUTS

Discrete PLC inputs and outputs can be thought of as electronic switches. An input is actually an optical isolator sensing either the presence or absence of a voltage. Likewise, an output is usually a triac or transistor which is turned on or off in accordance to the logical program running in the processor.

For this reason, it is extremely important to purchase the correct input and output modules for the types of inputs and outputs used in the application. For example, you should not attempt to use a 12-Vdc input module to sense the absence or presence of 120 Vac. Otherwise, the input module would probably be destroyed.

A small PLC designed for applications requiring only discrete I/O (courtesy of Eaton Corporation, Cutler-Hammer Products).

Likewise, if you attempted to use an ac-output module to control a dc-powered solenoid, the solenoid would probably never turn off after it was turned on the first time.

All PLC manufacturers have a wide variety of I/O modules designed for specific voltage and current requirements. The method of physically wiring the inputs and outputs to the PLC system varies greatly from one manufacturer to another. In most cases, some form of barrier strip or terminal strip is provided on the I/O modules or the I/O tracks to facilitate ease of installation.

During the PLC program development, an input is always represented as a relay contact. An output is always represented as a relay coil. When you write a program, the normal condition of an input relay contact will be its actual state during the absence of its associated input voltage on the I/O track. For example, if you assigned input address 1 to a normally-closed relay contact, it would open when you applied the correct operational voltage to that position on the I/O track. Likewise, during the absence of the voltage, the relay contact will close. (**Note:** Address 1 does not necessarily have to be an input. Addressing the I/O will be discussed later in this chapter.)

A TYPICAL PLC SYSTEM

It is important for you to understand that PLC operations and capabilities vary greatly from one manufacturer to another. A typical PLC has a maximum memory capacity of 4K dedicated to discrete (relay-ladder) logic and 4K of memory dedicated to data logic. Besides a battery backup to maintain the program in the CMOS memory in the event of a power failure, it also has internal nonvolatile EAROM (electrically alterable read only memory) to permanently store the finished program. There is a maximum capability of 256 discrete I/O (inputs/outputs) and 32 data I/O words (16 bits).

Internally, you can use up to 200 internal relay coils, 256 internal data registers (16 bits), and 32 timers/counters. You also have 5 time bases available (.1 second, 1 second, 10 seconds, 60 seconds, and 1 hour) and four logic function registers (logical AND, OR, EX-CLUSIVE OR, and INVERTER).

The following memory map will help you to fully understand how all of these PLC features are arranged within the processor section.

Discrete Functions	Data Functions
Addresses 1-256 reserved for discrete I/O	Registers 1-32 reserved for data I/O
Addresses 257-456 reserved for internal relay ladder logic	Registers 33-288 reserved for internal data logic
Addresses T01-T32 reserved for timer/counters	Registers 401-432 reserved for timer/counter presets
Addresses 600-604 reserved for (5) internal time bases	Registers 501-532 reserved for timer/counter current values

You can refer to this memory map as you examine the workings of the PLC system.

Discrete Relay Ladder Logic

Addresses 1-256 are reserved for the discrete I/O. Thus, there are up to 256 possible input/output combinations. In other words, you could use 200 inputs and 56 outputs or 4 inputs and 252 outputs or any other combination required for your application. You also have the option of using only a portion of the maximum I/O capability.

The physical I/O connections are not made directly to the processor section. These connections are made through the discrete I/O modules, which are plugged into an I/O track. This I/O track contains 8 plug-in connectors for accommodating up to 8 I/O modules. In this typical PLC system, you can assume that each I/O module contains 4 inputs or outputs. (You must determine how many discrete inputs and outputs are needed for the application and then purchase the I/O modules accordingly.) Since one discrete I/O track accommodates 8 I/O modules and each module contains 4 inputs or outputs, you are limited to 32 discrete inputs or outputs per I/O track.

The I/O track is connected to the processor section by means of a communications cable. According to the memory map, the maximum capability is 256 discrete inputs and outputs. This means you can use up to 8 separate discrete I/O tracks with a single processor (8 tracks × 32 I/O per track = 256).

If you use multiple I/O tracks with this system, you must

differentiate one track from another by means of a thumbwheel switch labeled 1 through 8. For example, if you connected an I/O track to the processor and adjusted the thumbwheel switch to the #1 position, you would utilize addresses 1-32 of the reserved 256 I/O addresses. If you changed the thumbwheel switch to the #2 position, you would utilize addresses 33-64.

In this manner, you can theoretically connect up to 8 I/O tracks to the single processor and utilize the entire 256 reserved addresses. However, for most applications, this is not necessary. Obviously, there is no reason to purchase additional discrete I/O tracks if they will not be needed for the application.

The following diagram shows a typical discrete I/O track layout. As stated previously, this track can accommodate up to 8 I/O modules (labeled MOD#1 through MOD#8). Each I/O module contains four separate inputs or outputs. However, these inputs or outputs are not mixed on the same module.

TYPICAL DISCRETE I/O TRACK

MOD #1	MOD #2	MOD #3	MOD #4	MOD #5	MOD #6	MOD #7	MOD #8
1	5	9	13	17	21	25	29
2	6	10	14	18	22	26	30
3	7	11	15	19	23	27	31
4	8	12	16	20	24	28	32

The numbers listed below the module numbers are the address locations that will be allocated by any I/O module you plug into a particular position on the I/O track.

Examine a simple rung of relay-ladder logic, and compare it to the previous I/O track layout.

```
---------|    |------------------------------------------------------------------(20)
         1
```

In this example, when you receive an input (voltage present) to the first position of the first I/O module, you will receive an output on the fourth position of the fifth I/O module. Upon examining the I/O track layout, you will find that address 1 is the physical position of the first I/O module and the first input on that module. This tells you that the first I/O module is an input module. If you plugged an output module into the first position of the I/O track, the previous rung of relay-ladder logic would not work. It also tells you that addresses 2, 3, and 4 must also be inputs because you cannot mix inputs and outputs on the same module.

The relay coil is labeled address 20. Since a coil is an output, you know that module 5 is an output module and that addresses 17, 18, and 19 are also outputs.

You have the option of plugging input or output modules into any position desired on the I/O track and the address of any individual input or output will depend on its physical position on the track. Also, you can mix I/O modules with different voltage and current ratings on the same I/O track.

Data Logic

With a typical system, a data I/O track cannot be used as a discrete I/O track. Instead, it must be plugged into a different connector (or port) on the processor. Also, the data I/O modules will only plug into a data I/O track.

As the memory map indicates, the first 32 data registers are dedicated to the data I/O. Each data I/O track will accept up to 8 data I/O modules. Each I/O module will utilize one data register. This means you can use up to 4 data I/O tracks with a single processor (4 tracks × 8 modules per track = 32 data registers).

As in the case of the discrete I/O tracks, the data I/O tracks also have a thumbwheel switch to define the track number. The following diagram illustrates the register locations which will be allocated by the data I/O track that has been defined as track #1.

TYPICAL DATA I/O TRACK

MOD #1	MOD #2	MOD #3	MOD #4	MOD #5	MOD #6	MOD #7	MOD #8
R001	R002	R003	R004	R005	R006	R007	R008

The #2 data I/O track would allocate registers 9 through 16, etc. The register associated with any individual data I/O module depends on the physical module location. For example, if a data I/O module is plugged into the fourth position of track #1, it will allocate register #4.

It is important for you to understand the concept of the data logic versus the discrete logic. The data logic is capable of performing logical decisions and arithmetic calculations on words. These words may represent analog voltage or current levels, counts, timer values, etc. Discrete logic only controls on-off type functions. You cannot read analog levels or large numbers with discrete logic.

For example, if you wanted to compare a process target value of 7 Vdc with the actual process value, you would have to use the data logic. If you wanted to turn on a motor contactor whenever a series of logical on-off conditions occurred, you would use the discrete logic.

In PLC systems containing data logic functions, the data logic works in conjunction with the discrete logic. For example, if you define a coil as address 260 in your discrete relay-ladder program, you can insert a set of normally open contacts in the beginning of a data line and label the contacts address 260. Whenever the relay-coil address 260 is energized in the discrete relay-logic program, the contacts labeled address 260 will close in the data program, allowing the function on that data line to be performed. Most data functions also have relay coils associated with them. These relay coils allow a data function to cause a logical action to occur with the discrete program. In this manner, discrete functions can control data functions and vice versa.

If you have never had any experience with a PLC, many of the above concepts will probably be difficult to understand. The following example describing an actual application (with its associated program) will illustrate how all of these PLC capabilities are actually incorporated.

IMPLEMENTATION OF A PLC CONTROL SYSTEM

Assume you have a batch chemical process you wish to control with the typical PLC system described previously. In this chemical process, you dump a chemical mixture into one of five separate heater tanks. This chemical mixture must reach a temperature of 190 °F before opening an output (flush) valve to allow the chemical to proceed to the next step in the process. Each heater tank must be completely emptied before you can allow a new batch of chemical to refill it.

Each of the five heater tanks has its own separate heater, temperature controller, and level sensor control. The temperature controller has an output that energizes the heater contactor until the internal chemical temperature reaches 190 °F. Upon reaching this temperature, the temperature controller drops out the heater contactor. The level sensor detects when the chemical level drops to a low enough level for the heater tank to be considered empty. When the tank is emptied, an output from the level sensor control automatically opens the input (refill) valve and closes the output (flush) valve. When the tank is completely filled, the level sensor control closes the input valve.

It takes approximately 6 minutes for each heater tank to empty. The output from all of the heater tanks eventually goes to one main pipeline. If the heater tanks are emptied sequentially, you can maintain an almost constant flow through the main pipeline. In other words, while one tank is being emptied into the main pipeline, the other four tanks are being heated. Since it takes approximately 15 minutes for the chemical to reach 190 degrees after the heater tank is refilled, you have plenty of time to heat a new batch while you are sequentially emptying the other four heater tanks. This process of sequentially emptying the heater tanks to maintain a constant flow through the main line is the function you want to automatically control with the PLC system.

You can begin by considering the PLC inputs needed. Basically, you must look at two variables from each heater tank: the chemical temperature and level. If the correct chemical temperature has been obtained and the level sensor control does not indicate the heater tank to be empty, you can empty the tank. However, you should not empty a heater tank if another heater tank is in the process of being emptied. You must empty the heater tanks sequentially to ensure that the chemical does not stay in any one tank (at 190 degrees) for an excessive amount of time.

As stated previously in this chapter, you begin by making an I/O listing. The following is an input listing for your hypothetical batch chemical heating application.

Tank #1 Temp confirm	Tank #1 Low-level
Tank #2 Temp confirm	Tank #2 Low-level
Tank #3 Temp confirm	Tank #3 Low-level
Tank #4 Temp confirm	Tank #4 Low-level
Tank #5 Temp confirm	Tank #5 Low-level

The "Temp confirm" signal from each of the tanks is obtained from a normally closed auxiliary contact from the heater contactor. Whenever the heater contactor drops out (meaning the correct temperature has been reached), the tank is ready to be emptied.

The "Low-level" signals result from an internal control relay contact in the level sensor control. This signal will only be present when the tank level is low enough to be considered empty.

The relay contacts in the level sensors are converted to "on/off" voltage levels by connecting them in series with 120 Vac. This allows you to use 120 Vac input modules for all of the discrete input signals. (If you can use the same type of input module for all of your inputs, you only need to carry one spare input module for replacement in the event of a module failure.)

Now that you have listed the inputs, you can define the inputs relative to their physical connection (and consequent memory address) on the discrete I/O track. Since you have a total of ten inputs, you need at least three input modules (4 inputs/module × 3 modules = 12 inputs total). The following table is an example of how you could define the inputs.

Tank #1 Temp confirm — Mod #1, Input 1 Address 1
Tank #2 Temp confirm — Mod #1, Input 2 Address 2
Tank #3 Temp confirm — Mod #1, Input 3 Address 3
Tank #4 Temp confirm — Mod #1, Input 4 Address 4
Tank #5 Temp confirm — Mod #2, Input 1 Address 5
Tank #1 Low-level — Mod #2, Input 2 Address 6
Tank #2 Low-level — Mod #2, Input 3 Address 7
Tank #3 Low-level — Mod #2, Input 4 Address 8
Tank #4 Low-level — Mod #3, Input 1 Address 9
Tank #5 Low-level — Mod #3, Input 2 Address 10
"Spare" input — Mod #3, Input 3 Address 11
"Spare" input — Mod #3, Input 4 Address 12

Remember, the address given to each input depends upon the input module position on the I/O track and the input used on that I/O module. (Refer to the typical discrete I/O track shown previously.)

Now that you have the inputs listed and defined, you can continue listing and defining the outputs in the same manner. The only outputs you need to control are the output (flush) valves for each tank. As stated previously, the input valve is already being automatically controlled by the level sensor control. The following table is an output listing and definition.

FV #1 open — Mod #4, Output 1 Address 13
FV #2 open — Mod #4, Output 2 Address 14

FV #3 open	— Mod #4, Output 3 Address 15
FV #4 open	— Mod #4, Output 4 Address 16
FV #5 open	— Mod #5, Output 1 Address 17
"Spare" output	— Mod #5, Output 2 Address 18
"Spare" output	— Mod #5, Output 3 Address 19
"Spare" output	— Mod #5, Output 4 Address 20

You will need a minimum of two output modules for five outputs (four outputs per module). This leaves you with three spare outputs.

Now that the entire I/O has been listed and defined, you can begin to write the program. Following is the discrete portion of the program. It consists of 7 rungs of relay-ladder logic. (A rung of relay-ladder logic is the associated logic desired to energize one coil.)

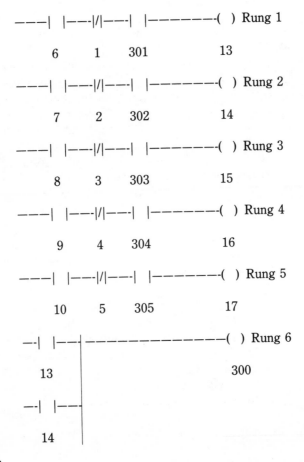

```
———|  |——-|/|——-|  |————————-( ) Rung 1
    6      1      301              13

———|  |——-|/|——-|  |————————-( ) Rung 2
    7      2      302              14

———|  |——-|/|——-|  |————————-( ) Rung 3
    8      3      303              15

———|  |——-|/|——-|  |————————-( ) Rung 4
    9      4      304              16

———|  |——-|/|——-|  |————————-( ) Rung 5
   10      5      305              17

 —-|  |——|————————————-( ) Rung 6
   13                            300

 —-|  |——|
   14
```

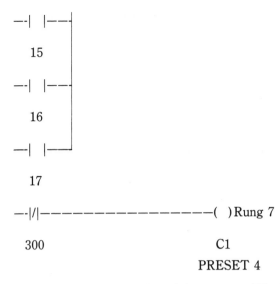

```
—| |—
  15
—| |—
  16
—| |—
  17

—|/|————————————————( ) Rung 7
 300                          C1
                          PRESET 4
```

This concludes the discrete portion of the program. The following is the data-logic program, which will work in conjunction with the discrete program to provide the desired results. (Lines of data logic cannot be referred to as rungs because they do not represent true relay-logic functions.)

```
———[R501] = [0000]—————————( ) Line 1
                                     301

———[R501] = [0001]—————————( ) Line 2
                                     302

———[R501] = [0002]—————————( ) Line 3
                                     303

———[R501] = [0003]—————————( ) Line 4
                                     304

———[R501] = [0004]—————————( ) Line 5
                                     305
```

NOTE: While reading the following operational description of this program, please refer to the I/O definition listing and the example PLC memory map.

The first five rungs of discrete logic are needed to energize one of the five flush valves. The first contact of each rung (addresses 6 through 10) is the input from each tank's level sensor control. These normally-open contacts (inputs) remain closed until the level in the associated tank drops low enough to be considered empty.

The second contact of each rung (addresses 1 through 5) is the input from each tank's heater contactor. These normally-closed contacts will remain open as long as their associated heater contactors are energized. When the heater contactors drop out (de-energize), the correct chemical temperature has been obtained, and the associated contacts revert to their normally-closed state. The coils on each rung are the flush valve outputs (addresses 13 through 17). As these coils are energized, their associated flush valves will open, allowing the heater tanks to empty.

The logic of rung 6 energizes the internal address 300 coil whenever one of the five flush valves is energized. This is accomplished by assigning addresses 13 through 17 to the parallel, normally-open contacts of rung 6. These addresses are the same as the output addresses in rungs 1 through 5. Therefore, whenever a flush valve is energized, the normally-closed address 300 contact in rung 7 will open.

Finally, the logic of rung 7 increments counter #1 by one count every time the normally-closed address 300 contact closes. (Opening the address 300 contact will not increment the counter.) Counter #1 is preset to return to zero (reset) when a count of 4 is obtained.

Upon examining the data-logic program, you will note that each line contains a register address (R501), which is being compared to a constant value. (When writing a data-logic program, you have the option of substituting a constant value in place of a register value. A constant is simply a number that never changes.)

Lines 1 through 5 of the data-logic program ensure a sequential emptying of each heater tank. As shown in the memory map, register 501 is the current value of counter #1. Upon system start-up, the current value of counter #1 will be zero. Therefore, the only data line that will be true under this condition will be line #1 (R500 = 0000). This true condition will cause the address 301 coil to energize. None of the other coils in the data-logic program will energize, because none of the other data lines will be true.

Now that you have learned the basics of the discrete and data-logic programs, the sequential program operation can be discussed from the beginning of the system start-up.

When you first turn on the system, all of the heater tanks are cold. This causes the tank heater contactors to energize. If all of the heater contactors are energized, all of the normally-closed "Temp confirm" contacts (addresses 1 through 5) are open. This means none of the output coils (addresses 13 through 17) can energize. This also keeps the address 300 coil in rung 6 from energizing. Counter #1 in rung 7 is currently at zero. Line 1 of the data logic is true (R501 = 0000). This causes the address 301 coil to energize. None of the heater tanks are empty, so the "Low-level" contacts (addresses 6 through 10) are closed. All of these conditions will remain stable until the heater contactor for heater tank #1 de-energizes (indicating that you have obtained the temperature target of 190 degrees).

Please refer to rung 1 of the discrete program. When the heater contactor for heater tank #1 de-energizes, the normally-closed address 1 contact closes. As stated previously, the true condition of data line 1 has energized the address 301 coil, causing the address 301 contact in rung 1 to close. The heater tank is not empty, so the address 6 contact is also closed. All of these conditions satisfy the logic conditions to energize address 13 output coil, which energizes the flush valve for heater tank #1.

In rung 6, the normally-open address 13 contacts have closed (because the address 13 output coil has energized in rung 1), causing the address 300 coil to energize. When the address 300 coil energizes, the normally-closed address 300 contacts in rung 7 open. This action does not change the current value of counter #1, because all counters in this typical PLC require a contact closure to increment. Since the current value of counter #1 has not changed, nothing has changed in the data-logic program.

When the level in heater tank #1 decreases to the empty level, the normally-open address 6 contact will open, causing the address 13 output coil to de-energize. This action de-energizes the flush valve for heater tank #1, causing the flush valve to close.

When the address 13 output coil de-energizes, its associated normally-open address 13 contact in rung 6 opens. This de-energizes the address 300 coil (in rung 6) and closes the normally-closed address 300 contacts in rung 7. Counter #1 increments to a current value of one.

Since the current value of counter #1 has changed to 1, line

1 of the data-logic program is no longer true. As a result, the address 301 coil de-energizes. Now, data line 2 is true (R501 = 0001), causing its associated address 302 coil to energize. Going back to the discrete logic, this action causes the normally-open address 302 contact in rung 2 to close. Assuming heater tank #2 is full and at the correct temperature, you should have satisfied the logic conditions to energize the address 14 output coil. This energizes the heater tank #2 flush valve, and heater tank #2 begins to empty.

From this point, the logic continues in the same manner as it did after the heater tank #1 flush valve was energized. As heater tanks #2 through #5 are sequentially emptied, heater tank #1 is automatically refilled (by means of its own internal logic), and its fresh chemical batch is heated to the target temperature.

When heater tank #5 is empty, counter #1 will reset back to zero (because it has a preset value of 4), and the whole process will begin again.

WHY USE A PLC FOR THE PREVIOUS APPLICATION?

If you are familiar with various automation techniques, you may be wondering why a PLC system should be used for this application. Numerous factors must be considered before this question can be answered.

If you know your logic is correct and you know the system requirements will never change, it may be more cost-effective to purchase discrete relays and counter hardware (or a stepper switch) and achieve the same results. (A discrete system is sometimes referred to as a *hard-wired system*.) However, in most cases, there may be at least one unknown out of the two mentioned previously.

Assume you install the PLC system exactly as described previously. It operates perfectly for two weeks until you experience a temperature-controller problem with heater tank #2. The temperature controller must be removed from the tank for repair. In the meantime, you discover the PLC system will no longer operate because it is waiting to empty heater tank #2, but the level sensor is indicating the tank is already empty (in reality, it is) and you cannot fill the tank because you cannot heat the chemical.

With a hard-wired system, you must perform a temporary wiring change until you repair the temperature controller. With the PLC system, you can simply make the following modification to rung 2 of the discrete logic program:

```
————————| |——|/|—|—| |————————————( )  Rung 2
          7    2  │ 302              14
————————|/|————┘
```

This will produce a very fast pulse (scan time of the PLC) when-ever the address 302 contact closes. The fast pulse will increment counter #1 which will cause the program to quickly bypass the #2 heater tank. This program modification would require much less time than a wiring modification and can be easily removed when the temperature controller is repaired.

Consider another possibility. Suppose you wish to add one (or more) heater tanks to the present system. This can be accommo-dated very easily with a PLC system.

Over a period of time, PLC systems tend to grow as you gain experience, efficiency, and production. For example, suppose you wish to improve the operation of the whole system by monitoring the actual flow through the main pipeline (using an appropriate flow-meter) and controlling the flow with a proportional valve placed in-line with the main pipeline. This can be accomplished very eas-ily through the data I/O by reading the flowmeter signal and out-putting a control voltage to the proportional valve. In addition to the flowmeter and proportional valve, you would have to purchase a data I/O track, an analog-input data module, and an analog-output data module for this purpose.

Once you have a digital word representing the actual flow, there are numerous ways you can control it. Many PLCs with data han-dling capability already have simulated PID loops in the ROM pro-gram. If this is the case, you simply define which PID loop you wish to use and insert the PID variables into the program. You can achieve similar results without PIDs by using the mathematical functions available to you in the data program.

If you implement a proportional-control system as described previously, you will have all of the data necessary to document the production throughput for this process. For example, you may wish to print out a hard-copy report showing the total number of gal-lons of chemical which has flowed through the main pipeline dur-ing each eight-hour shift. From a hardware point of view, you only need to purchase a printer interface from the PLC manufacturer and a compatible printer. You will probably be capable of making the required program modifications. The PLC stores the data to be printed in one of many shift registers or data tables available in the data program.

In other words, the sky is the limit. A PLC system offers you the flexibility to experiment with and modify the program to meet your changing process needs or improve your competitive edge. In addition, you can write the programs in a language you already understand or can learn very easily. In most cases, the operator manuals supplied with the PLC system can be read and easily understood by anyone with basic electrical knowledge.

TROUBLESHOOTING PLC SYSTEMS

Virtually all PLC systems incorporate some form of internal diagnostic programming. The PLC user's manual will provide specific information regarding the use of its unique diagnostic routine for troubleshooting purposes. In most cases, PLCs are very easy to troubleshoot.

FINAL CONSIDERATIONS

The PLC system you have just examined is typical. Since PLC systems vary greatly from one manufacturer to another, you must choose a system with the capabilities necessary for the application without the unnecessary cost of overkill.

Some manufacturers claim their small shoebox PLCs can cost-effectively replace as few as 10 discrete relays. These shoebox PLCs have limited I/O capability and do not offer any data-handling functions. Although they are an excellent alternative to discrete relay logic in some applications, their limited capabilities can cause problems in many installations. For example, you could have sequentially emptied the heater tanks (in the previous hypothetical chemical batch process) with a software stepping switch (or drum controller) using a small PLC system. Unfortunately, if you wanted to monitor and control the main pipeline flow at a later date, it would be impossible without data programmability.

If you wish to obtain a better understanding of PLCs than is possible in a general writing of this nature, you can purchase the user's manual for a medium-sized PLC from one of many PLC manufacturers. The user's manual will provide specific information and programming examples in addition to listing optional equipment and application data.

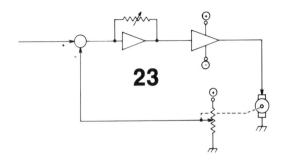

23

Photoelectronics

S OME OF THE MOST EXCITING ADVANCEMENTS IN THE INDUS-
trial-control field has been in the area of *photoelectronics*. Photoelectronics can sense the presence of any object without contacting the object. But photoelectronic systems have the severe
disadvantage of being sensitive to ambient light and contamination.
Also, their reliability is not satisfactory in some applications due
to the need for frequent sensitivity adjustments to compensate for
the aging of the filaments in the light sources used. The light
sources themselves (usually incandescent lamps) are also incapa-
ble of withstanding high vibration and have a limited life span. See
Fig. 23-1.

In order to understand how the above problems have been resolved, you must first study the workings of a typical photoelectronic system (also called a "photoeye" or "scanner"), as well as
some of the terms associated with photoelectronics.

TYPES OF PHOTOELECTRONIC SYSTEMS

Basically, there are only two main classifications of photoeye
systems: *transmission* and *retroreflective*.

A transmission photoeye system consists of an emitter (light
source) and a receiver (light sensor) that are mounted in such a way
that they oppose each other. In other words, the emitter transmits
a beam of light directly to the receiver. Any object interrupting this

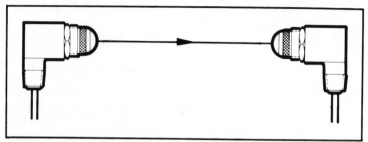

Fig. 23-1. Illustration of transmission (thru-beam) scanning (courtesy of Micro Switch, a Honeywell Division).

beam will be detected by the receiver.

A retroreflective photoeye system also contains an emitter and receiver. The emitter is mounted so that the beam of light is intentionally reflected from a highly reflective material (such as a bicycle-type reflector or reflective tape) and the reflected beam is received by the receiver. A retroreflective system creates two beams of light: the original beam from the emitter and the reflected beam from the reflector. Any object interrupting either beam will be detected by the receiver. See Fig. 23-2.

In simple terms, the receiver sees the beam directly in a trans-

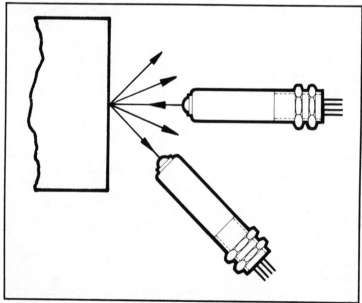

Fig. 23-2. Illustration of retroreflective scanning (courtesy of Micro Switch, a Honeywell Division).

mission photoeye system. In a retroreflective photoeye system, the receiver sees a reflected beam.

The primary advantage of the transmission photoeye system is that the receiver sees the beam directly. Therefore, the beam is stronger and more immune to dirt or contamination buildup on the internal optics of the eye. Also, the distance between the emitter and receiver, called the "range," can be greater than that of a retroreflective system.

A retroreflective system can be designed so that the beam is reflected almost directly back to the emitter. For this reason, the receiver can be mounted in the same physical housing as the emitter. This one-piece construction has the advantage of being easier to mount and align. However, care must be exercised that the object breaking the beam is not reflective.

Another type of retroreflective photoelectronic method is called *diffused-proximity scanning* (sometimes referred to as *reflective scanning*). The light emitted from a diffused-proximity scanner is diffused (scattered) in such a way that it creates a light cloud close to the scanner housing. Virtually any object coming within the area of this light cloud will reflect enough light back to the receiver to be sensed (the object does not have to be reflective). In other words, in a diffused-proximity system, the object to be sensed actually makes the beam instead of breaking it. From an applications point of view, a diffused-proximity scanner is analogous to inductive- or capacitive-proximity switches. See Fig. 23-3.

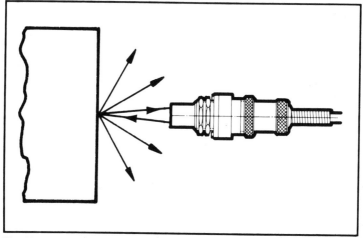

Fig. 23-3. Illustration of diffused proximity scanning (courtesy of Micro Switch, a Honeywell Division).

A *convergent-beam photoelectronic system* is similar to diffused-proximity scanning. However, instead of scattering the beam from the emitter, convergent beam systems incorporate optics designed to focus the beam at a fixed distance from the front surface of the scanner housing. The effect of focusing the beam gives a convergent-beam scanner the unique ability to only sense the presence of an object at a precise distance from the scanner housing. Unlike diffused proximity scanning, a convergent beam scanner will not sense an object if it is too close or too far away.

Internal scanner optics will tailor the performance of the photoelectronic system. Factors such as range (operating distance), diffusion (how the beam is scattered), effective beam, excess gain, contrast, and system alignment vary with each type of internal optical system.

Depending on the type of optics used in an emitter, the light being produced by the emitter will either spread out (in transmission and retroreflective systems), converge to a point and then spread out (in convergent systems), or diffuse (in diffused-proximity system). In any case, the "effective beam" is the part of the beam that must be affected to cause reliable "sensing" action from the photoelectronic system. The effective beam should not be confused with the actual *beam pattern*. In theory, the light-radiation pattern from the emitter of a transmission system will spread out to infinity just as light radiation spreads out from a hand-held flashlight. The area of the beam pattern that spreads outside of the optical lens area of the receiver is obviously not "seen" by the receiver. Therefore, interrupting this "wasted" part of the beam pattern will have no effect on the receiver and is not considered part of the effective beam.

Excess gain is the amount of "excess" light falling on a photoelectronic receiver above the minimum amount of light required to operate the receiver at its minimum threshold level. In modern photoelectronic systems, depending on range and alignment, excess gain levels of 1000 × or higher are possible. An excess gain of 1000 × means the light intensity falling on the receiver is 1000 times more powerful than it needs to be for receiver detection.

One final term to become familiar with is *contrast*. In many applications, the object to be sensed by the photoelectronic system may present one of two problems: it may be somewhat translucent, or it may block less than 100 percent of the effective beam. In either case, the light radiation being received by the receiver never drops to a zero level. Contrast defines the ratio between the "light"

Examples of fiber-optic scanners (courtesy of Micro Switch, a Honeywell Division).

condition and the "dark" condition. For this reason, contrast is often referred to as the "light-to-dark ratio." Contrast is calculated by dividing the excess gain during the "light" condition (effective beam unblocked) by the excess gain during the "dark" condition (effective beam blocked). For example, if a photoelectronic system showed an excess gain of 10 × during the light condition and 0.5 × during the dark condition, the contrast would be:

$$\frac{10X}{0.5X} = 20$$

The sensitivity adjustment on a photoelectronic system compensates for contrast problems provided that the contrast ratio is at least 3 or above. Contrast levels below 3 are considered unreliable, and alternate sensing methods should be researched.

THE NEW GENERATION OF PHOTOELECTRONIC CONTROLS

Virtually all of the problems associated with the older generation of photoelectric controls have been eliminated in the newer generation of photoelectronic devices. The problems associated with the light sources were resolved by using light-emitting diodes instead of incandescent lamps. LEDs have an almost unlimited life span, are not affected by vibration, produce virtually no heat, and do not degrade over a period of time.

The problem of sensitivity to ambient light is eliminated by modulating the LED light source. This simply means that the LED is turned on and off (pulsed) at a high rate of speed (typically 7 kHz-20 kHz). This is accomplished by an oscillator circuit in the emitter section. The receiver section incorporates a bandpass filter (a filter which only allows a narrow range of frequencies to pass) that is tuned to the exact frequency of the emitter modulation. For this reason, the modulated light beam from the emitter is easily seen by the receiver, but ambient light is completely blocked. In other words, ambient light would have to be modulated at the same frequency as the emitter beam before the receiver could detect it. This principle, in nature, is similar to the way that a radio receiver tunes solidly to one station while ignoring all the other radio waves that may be present in the room.

The insensitivity to ambient light is not the only advantage of modulated photoeye systems. It is possible to pulse a LED at a much higher energy level than can be held at a continuous level (without destroying the LED). This high-energy level is what gives a modern photoeye system the ability to penetrate (burn through) contamination on the optical system. Sensitivity to alignment errors is also reduced. For this reason, modern photoelectronic systems are much easier to install than their incandescent predecessors. (The light radiation burn-through can also be used for special control applications, which will be discussed in a later chapter.)

Although there are many advantages to modulated photoelectronic systems, there are some special applications where nonmodulated photoelectronic systems (incorporating LEDs) may be used. Basically, nonmodulated systems may be used for any application requiring a very short scanning range and where there is no concern with dirt accumulation, ambient light, or penetrating power. Another possibility you may encounter is an application in which you need to sense objects that emit their own light. For example, red-hot metal and glass have a very high infrared light emission. As long as this infrared light emission is much higher than the ambient light level, a nonmodulated receiver may be used to sense them. An emitter is not needed, because the objects themselves become individual emitters.

Before ending the subject of modulated verses nonmodulated photoelectronic systems, I would like to comment on a very common misunderstanding.

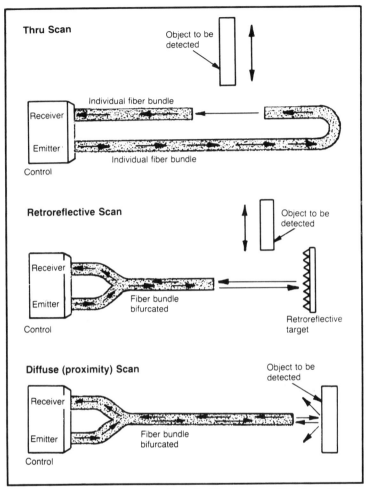

Fig. 23-4. An example of modern photoelectric scanners. The above illustration includes transmission, diffused proximity, and retroreflective types (courtesy of Eaton Corporation, Cutler-Hammer Products).

Care should be taken when replacing nonmodulated incandescent photoelectronic systems with nonmodulated LED systems. The illuminating power of an LED is less than 1/10th that of a small incandescent lamp.

ALIGNMENT

Alignment of some photoeye control systems may be somewhat difficult. Although sensitivity to alignment errors is reduced in

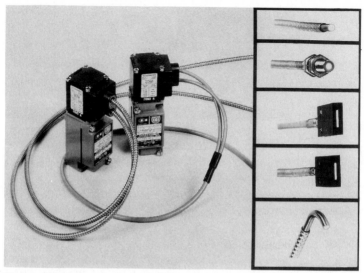

Modern fiber-optic sensors include transmission (thru-beam) and bifurcated bundles for proximity sensing. Examples of different tip configurations can handle unique applications (courtesy of Eaton Corporation, Cutler-Hammer Products).

modulated LED photoelectronic systems, it is still critical when the distance between the emitter and receiver approaches 200 feet. For this reason, some photoelectronic manufacturers incorporate electronic alignment devices for ease of system installation.

For example, one manufacturer installs a flashing LED indicator on the top of the receiver. The flashing "rate" of this LED is proportional to the excess gain of the system. If a poor alignment exists, the LED will flash at a very slow rate. As alignment is improved, the flash rate will increase until it cannot be seen by the human eye. At this point, alignment is correct. Another large manufacturer uses multicolored LEDs to aid in alignment.

IR AND VISIBLE LIGHT

Because there are many different applications for modern photoeye systems, most manufacturers offer both IR (infrared) and visible light beams. Visible light photoeye systems are easier to align because the beam is visible. They are also color-sensitive. Sensitivity to color is not necessarily a disadvantage. For example, when scanning bar codes (such as UPC labels), color differentiation is a necessity.

Infrared systems are used when sensitivity to color is a disadvantage. Also, the sensitivity of the receiver is usually higher because of the physics involved with the solid-state receiving devices (phototransistors and photodiodes).

FIBER OPTICS

Fiber optics have the unique ability to transmit light around corners. They are very useful for extremely small and precise applications and can be configured inexpensively to detect unusual shapes by changing the size or shape of the fiber bundle ends. Fiber-optic cables can be used in environments in which any electronic equipment capable of producing a spark is not permitted. Fiber optics can also be used in high-temperature or adverse environments that would destroy standard photoeyes.

A fiber-optic cable consists of many small transparent plastic (or glass) strands. The bundle is housed in a flexible sheath, and the ends of the cable are shaped according to the application. See Fig. 23-4.

It is important to understand that the utilization of fiber optics only changes the media through which the effective beam is transmitted. The optical tailoring is accomplished by shaping the ends of the bundles and by intermixing (bifurcating) the emitter and receiver strands.

A fiber-optic photoelectronic system consists of an emitter and receiver as do the photoelectronic systems previously discussed. If two fiber-optic cables are used, one cable will be the emitter light-pipe, and the other will be the receiver light-pipe. Fiber-optic proximity scanners are formed by intermixing the emitter and receiver strands into a single bifurcated bundle. Bifurcated cables are used in register-mark sensing and unique parts orientation and/or sensing.

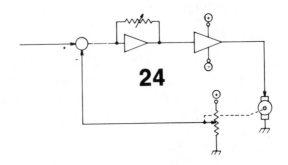

Noncontacting
Sensor Applications

T HE FOLLOWING EXAMPLES ARE INTENDED TO DEMONSTRATE
how noncontacting sensors (photoelectronic, inductive proximity, and all-metals proximity) can be applied to industrial control systems. These examples are nonproprietary and general.
Therefore, you should determine the specifics involved in your particular application (scanning distance, environmental characteristics, load-voltage requirements, etc.).

When you design the logic utilized in photoelectric or proximity sensing, you should always try to incorporate a failsafe. This
means, under normal operating conditions, a relay or solid-state
output device is maintained in an energized state and drops out
when an object is seen. The reasoning behind this philosophy is
simple. In the failsafe mode, if the power is interrupted or the output device fails, there will be a constant indication that a problem
exists. If a failsafe is not incorporated in the logic, it is harder to
detect an existing problem which may eventually cause serious
equipment or property damage.

When you consider the cost of noncontacting systems, you
should keep a realistic perspective on the initial cost as well as the
installation cost. In many cases, the installation cost will far exceed the initial systems cost. Following are a few rule-of-thumb
generalities you can consider:

1. In most cases, a modulated, infrared, transmission-

photoeye system is the best performing photoelectronic system. Its initial cost and installation cost are usually much higher than other systems. This is because you have to install two eyes (emitter and receiver) instead of one.

2. Inductive proximity sensors are just as reliable and much less expensive than all-metals proximity sensors. This is true only if the application requires the sensing of a ferrous metal material.

It should also be mentioned that there are literally thousands of different applications which are not shown in the following examples. See Fig. 24-1. If you should have any questions regarding applications which do not closely resemble those found in this book, most sensor manufacturers have qualified application engineers who will be very happy to help you.

PHOTOELECTRONIC APPLICATIONS

NOTE: The author does not recommend the use of commercially available photoelectronic or proximity systems in applications where human life may be endangered by their failure. See Fig. 24-1.

Fork Lift Load Height Protection

There have been many cases when a loaded fork lift, trying to enter a doorway, has lifted its load higher than the doorway thus causing a collision. To overcome this problem, a method is needed to warn the operator if an object was too high to pass through a doorway.

For this application, a modulated, infrared, transmission photoeye system with a 20-foot range is chosen. The output device is an on-delay relay with an adjustable time period of up to 30 seconds. As a means of providing the failsafe, the system is placed in the light operate mode and the warning light (and buzzer) operates from the normally closed contacts of the relay. The photoeyes are opposing each other and placed at the maximum allowable height. The photoeyes are also placed at a sufficient distance in front of the doorway to allow the operator time to stop if the maximum allowable height is exceeded.

How it Works. Since the system is in the light-operate mode, under normal conditions, the beam will always be seen by the receiver. Therefore, the on-delay TDR will be maintained in the ener-

Typical applications

The drawings here give you only a hint of the variety and potential for photoelectric control applications.

A. Two light source-photoreceiver pairs are used to keep hopper fill level between high and low limits.

B. Counting products is a common application of photoelectric controls. Counting batches or groups of cans or other items prior to packaging or group processing is also common.

B.

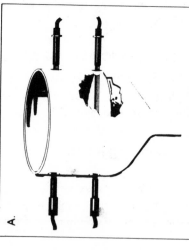

A.

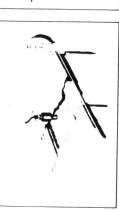

C.

D.

E.

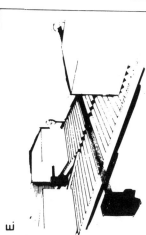

F.

C. A photoelectric control operating on reflected light is a simple way to detect a web break. An alternative is to put a light source above the web, and a photoreceiver below.

D. Dark caps are checked for white liners by a photoelectric scanner. The scanner activates a mechanism that rejects caps without the liners.

E. To prevent collisions where two conveyors merge, each conveyor is monitored by a control that powers the other conveyor when its own is cleared.

F. A tubular light source and photoreceiver in a specially designed bracket detect registration marks to initiate any related operation, such as printing, cutoff, or folding.

Fig. 24-1. Photoelectric control applications (courtesy of Micro Switch, a Honeywell Division). Continued on page 272 and 273.

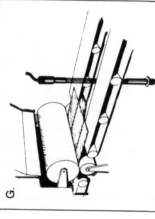

G.

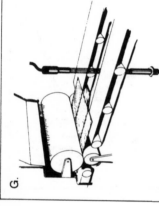

H.

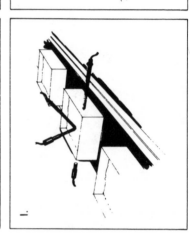

J.

I.

G. Gluing, buffing, or flattening can be done efficiently by controlling the pressure rollers or buffer with a photoelectric light source and photoreceiver that detect the product to be processed.

H. Using logic for one-shot pulse output, a photoelectric control slows a conveyor and fills the carton which has interrupted the light beam.

I. Two light source-photoreceiver pairs work together to check fill level. The box-detecting pair turns on, or enables, the fill inspection pair — thereby preventing the inspection pair from mistaking the space between boxes as an "improper fill".

J. Light source and photoreceiver placed near a guillotine are used to detect products and operate the blade for cutting the link between products.

K.

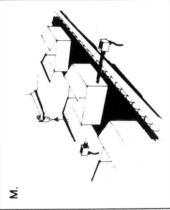

K.

M.

L.

K. Thread break detection is easy when the photoelectric beam is interrupted by a lightweight flag riding on the taut thread.

L. The size of a paper or fabric roll can be controlled by positioning a light source and a photoreceiver so the roll diameter blocks the beam.

M. Turning a glue nozzle on and off is an example of process control easily effected with a photoelectric control.

gized state. If a tall object comes between the photoeyes, the beam will be broken and the TDR will drop out. Its normally closed contacts will close causing the warning light and buzzer to activate. They will stay activated, even if the tall object is no longer present, for the on-delay setting of the TDR (max. 30 seconds).

Similar Applications. The same principle of sensing excessive height can be applied to sensing excessive width, overfill, unauthorized entry, multiple-part detection and many other non-life-threatening warning applications.

Trash Compactor Level Detection

This is an industrial trash compactor intended primarily for dry waste. When the compactor is full the operator presses a momentary start pushbutton to activate the compactor cycle. The compactor will perform one compact cycle and return to standby mode. Occasionally, plant personnel will neglect to press the start button when the compactor is full causing debris to spill on the floor. Therefore a method is needed to automate the trash compactor so that a preset trash level is detected and the compact cycle automatically starts.

This application is more convenient than critical. Therefore, a retroreflective system is chosen for cost advantage and ease of installation. The reflector is mounted on the back wall of the trash container at the proper height at which you want to initiate the compactor cycle. The photoeye is mounted on the opposite wall and is placed in the dark-operate mode. The output device is an on-delay relay with a time period of approximately 10 seconds. The normally open contacts of the TDR are placed in parallel with the start pushbutton contacts.

How it Works. The retroreflective photoeye sees the reflector until the trash level exceeds the height of the reflector. Since the photoeye is in the dark-operate mode, it tries to activate the TDR as soon as the reflector is blocked. The TDR requires the blockage to exist for at least 10 seconds before initiating the compactor cycle. This is to eliminate unnecessary compactor cycles from starting whenever trash is dumped into the trash container and only breaks the beam for a short period of time.

This system is not maintained in a failsafe mode due to the non-critical nature of this application.

Similar Applications. Use this system with any kind of non-critical container overfill indicator.

274

Nonmetallic "Part-in-Place" Confirmation

A non-metallic plastic material is placed in the center of a press. To improve the press' efficiency and to eliminate possible die damage, the presence of the plastic material should be sensed before allowing the press to cycle.

There are two methods commonly used for sensing the plastic material. The preferred method is to use a modulated, infrared, transmission-photoeye system to look through the press gap area for the plastic material. Unfortunately, due to the construction of many presses, this is not possible. The alternate method is to use an infrared, convergent-beam photoeye.

The convergent-beam photoeye is adjusted to see the material when it is in place rather than the absence of the beam if the material is in place. This is for the purpose of eliminating possible erroneous press cycling if the photoeye lens assembly becomes contaminated.

The output logic of either system is simple on-off logic in series with the start cycle pushbutton to enable a press cycle only if the material presence is sensed.

A failsafe is incorporated in either system by using the normally open contacts of the output relay. The transmission system is placed in the dark-operate mode and the convergent beam system is placed in the light-operate mode. In either case, if the photoeye fails, the output relay will not be energized and the press operation will not be enabled.

Paint Spray Booth Automation

Random objects of different shapes and sizes are being carried along a monorail to be painted upon arrival at the paint spray booth. A method is needed to detect the presence of an object and start an automatic spray painting operation.

This application is more complex than it appears. Due to the overspray contamination, it is impossible to place any kind of photoelectric system in close proximity to the paint spray booth. For this reason, the photoeye must be mounted before the paint spray booth to detect the part before it actually reaches the booth. You cannot use a TDR to compensate for the delay in the part arrival to the booth. The monorail is capable of changing speed so the time delay of the part arrival will not remain constant.

How it Works. A modulated, infrared, transmission photoeye system is chosen to sense the presence of the object to be painted on the monorail. This photoeye is placed approximately 15 feet before the paint spray booth. The photoeye is adjusted so that it will not see the monorail hooks but will see any object placed on them. The photoeye output relay is configured to latch whenever energized. When an object to be painted is seen by the photoeye, the output relay will latch and remain latched until acted on by an external device.

When the photoeye output relay latches, it enables a preset counter to start counting pulses being generated by an incremental encoder. The encoder shaft is connected to a rubber wheel riding on the monorail ways. Each pulse from the encoder is directly proportional to an exact distance of linear travel. Assume you calculated this distance to be equal to two inches. Since the photoeye is located approximately 15 feet before the spray paint booth, the preset counter is set at 90 to start the spraying operation (15 ft. $\times$ 12 = 180 in. of travel. 180/2 inches per pulse = 90). When the preset counter counts 90 pulses from the encoder, it outputs a pulse to start the spraying operation. This same pulse resets the counter back to zero and unlatches the output relay from the photoeye system.

Similar Applications. It should be evident that you were actually measuring distance in the above example. Any application requiring the measurement (length) of an object on a conveyor or distance traveled can be done in the same manner.

Boiler Protection: Coal Feed Blockage

As coal fuel passes through numerous crusher stages, a method is needed to detect if a blockage occurs which inhibits the flow of coal to the boiler feed chute.

A modulated, infrared, transmission photoeye is perfect for this application. (A high-power eye is recommended for this application.) Two clear acrylic windows are installed in the coal feed chute so the photoeye system can look through the chute. The photoeye is placed in the dark-operate mode. The photoeye sensitivity is adjusted high to facilitate burn-through of heavy contamination. The blockage alarm is initiated from the normally closed contacts of the output relay. The output relay is an off-delay TDR with a time period of approximately 10 seconds.

How it Works. During normal operation, the feeder chute is full of coal being fed into the boiler. As long as coal is present, the beam is broken keeping the output relay energized. If a blockage occurs, the beam is made and the output from the photoeye tries to de-energize the output relay. If the blockage is temporary (common to coal feeders) the off delay relay will not have a chance to drop out before the photoeye beam is again broken by coal in the chute. If the photoeye beam remains, the output TDR will time out after 10 seconds and initiate an alarm. The failsafe is incorporated in this application by maintaining the output relay energized during normal operation.

Similar Applications. This type of system can be used with solid material feeders of all kinds. Plastic pellet hopper low-level detection on plastic extruders. Solid materials fill level detection.

Automatic Conveyor Stop

A conveyor carries objects of varying shapes and sizes to a packaging room. An operator unloads the objects when they reach the end of the conveyor line and places them in boxes for packing prior to shipment. During periods of heavy production, a method is needed to automatically stop the conveyor to prevent the object from dropping off the end if the operator is too busy to unload it in time.

A modulated, infrared, transmission system with simple on-off logic is used. The photoeye is placed at the end of the conveyor so an object will break the beam before it falls off the end. The photoeye is placed in the light-operate mode. The conveyor run enable line is connected to the normally open contact of the output relay.

How it Works. Since the photoeye is placed in the light-operate mode, the output relay will energize whenever the beam is seen by the receiver. When the output relay energizes, the normally open contacts close causing the conveyor to run. If an object breaks the beam (meaning it's at the end of the conveyor), the output relay will de-energize stopping the conveyor until the object is removed by the operator. As soon as the beam is seen again by the receiver, the conveyor will begin moving.

Similar Applications. This method may be used for any kind of non-hazardous conveyor, automatic stop-start operation.

Packaging Machine Length Detection

As objects of various lengths are moved down a conveyor line, packaging personnel wrap the objects in a packing material. To improve the efficiency of the wrapping operation, the packaging personnel have to know how much packing material to pull off the roll for each item to be wrapped.

Two modulated, infrared, transmission photoeyes are used in this application. You also need a reset counter and an incremental encoder.

One photoeye system is placed at the entry of the wrapping station. This is the eye being used to measure the length of the object to be wrapped by enabling the reset counter to count the pulses from the incremental encoder. The other photoeye is placed to look across the wrapping station and is used to reset the counter. The incremental encoder shaft is mechanically connected to the conveyor drive with an appropriate gear reducer to provide one pulse per linear inch of travel by the conveyor.

How it Works. As the object to be wrapped approaches the wrapping station, it breaks the beam of the first photoeye system. When this happens, the photoeye enables the counter to begin counting the pulses from the incremental encoder. The counter will count as long as the beam from the first photoeye is broken. When the trailing edge of the object to be wrapped passes by the first photoeye, it inhibits the counter from counting any further pulses from the encoder. The counter is now displaying (on its display) the accurate length of the object (in inches) to be wrapped.

By this time, the object is in the wrapping station and breaking the beam of the second photoeye system. The wrapping personnel look at the counter display and tear off the proper length of wrapping material. The object is now removed from the wrapping station and placed on a wrapping table. As the object is removed from the wrapping station, the beam is again restored to the second photoeye system. This photoeye resets the counter to zero and the whole process starts over again.

Similar Applications. This method may be used for any conveyor line length measurement.

Pill Counter

A method is needed to accurately count vitamin pills from a large feeder bin as they are being placed in bottles for shipment. The count must be accurate. The high production rate requires multiple bottles to be filled at the same time.

A multiple fill station capable of filling 12 bottles at the same time was chosen. The mechanical operation of the fill station will be described first.

At the bottom of the feed bin is a rotating disc with 12 holes in it. The holes are sized so only one pill goes through at a time. The holes are equally spaced around the inside circumference of the disc. Directly below the disc are 12 chambers arranged in a circle. The width of these chambers is such that as the disc rotates each chamber can only be exposed to one hole at a time. At the bottom of each chamber is a feed tube with a small positive air pressure applied to force the pill through the tube. The tubes run down to the empty bottles.

As the disc rotates, the pills randomly fall through the holes one at a time into the chambers below. Each chamber receives one pill at a time due to the spacing of the holes in the disc.

Now that you have an understanding of the mechanical operation, the electronic operation can be discussed.

Twelve photoelectronic batch totalizing systems are incorporated for this application. Each system utilizes a modulated, infrared, transmission photoeye to look through a clear insert in each fill tube located just prior to the bottle top. The output of each photoeye is applied to an electronic preset counter with a one-shot output when the preset number is reached. This same one-shot output also resets the counter. The preset number is the number of pills needed to fill each bottle.

How it Works. As the pills are fed into the bottles, they are sensed by the photoeye system and counted by the preset counter. When the preset count is reached, the output from the counter momentarily closes a valve in the feed tube to stop any pills from dropping during a bottle change. This output also resets the counter and begins the bottle change. When an empty bottle is in place, the feed tube valve opens and the cycle is repeated.

Similar Applications. This method may be used in any high-speed batch totalizing application.

Blown Film Bubble Size Detection

A method is needed to detect when the bubble size of a blown film machine is the correct diameter. When the correct diameter has been reached, it should be maintained.

A set of modulated, infrared, convergent-beam photoeyes with simple on-off logic is chosen for this application. The first photoeye is positioned nearest to the bubble and placed in the light-operate mode. The second photoeye is positioned further from the bubble and also placed in the light-operate mode. A manual/auto switch is provided to allow the operator to override the automatic operation of the photoeyes.

The normally open contact of the first photoeye is used to operate the *increase* air pressure pushbutton on the air pressure control panel. The normally open contact of the second eye is used to control the *decrease* pushbutton from the same panel.

How it Works. The operator manually controls the air pressure when blowing a new bubble. Once the bubble is large enough to be seen by the first photoeye, the operator can then place the manual/auto selector switch in the auto position. The first photoeye will increase the bubble air pressure until the bubble surface is too close to be seen. (The bubble surface gets closer to the eye because the bubble diameter will increase as the air pressure increases.) As this equilibrium point is reached, all control action ceases. If the bubble diameter becomes too large, the bubble surface will come into the range of the second photoeye which will decrease the air pressure and reduce the bubble size.

Similar Applications. This method is used for any typical high-low level control or in-out positioning systems with either wide hysteresis and/or slow response. Utilizing this type of control in a system with a fast response or narrow deadband (hysteresis) would result in process oscillations.

Packaging Machine Fill Positioner

Round opaque plastic containers are placed in line on a conveyor to be moved under a fill station. The conveyor must be stopped when the container is directly under the fill head and a signal given which starts the filling process. There is no space between the containers.

A modulated, infrared, convergent beam photoeye with an adjustable one-shot logic is chosen for this application. It is mounted directly in-line with the fill head looking at the side of the containers as they pass. The photoeye is placed in the light-operate mode. The normally open contacts from the photoeye output relay are used to stop the conveyor and start the filling process.

How it Works. Since the photoeye is seeing the side of the round containers, the point nearest the photoeye will be the approximate middle of the container. The photoeye distance from the containers is carefully adjusted so that it only sees this point. When the container middle point is detected, the one-shot output energizes the output relay long enough to stop the conveyor and start the filling process. Because the photoeye is in-line with the fill head, the container will be directly under the fill head at this time. The one-shot output logic causes the photoeye output relay to drop out before the filling process is complete. This is necessary to allow the conveyor to restart when the container is filled.

NOTE: You cannot use a transmission sensing method for this application because there was no spacing between the containers.

Similar Applications. This application can be used wherever the sensing of the widest point for positioning or error detection (as in a wrong part ejector or wrong orientation detection) is needed.

Bottle Cap Present

A method is needed to detect uncapped bottles and reject them. The bottle caps are very flat so their presence cannot be detected by a differential in overall bottle height.

You must use two photoeye systems for an application of this type. The first system will form an interrogation window to tell the second photoeye when to look for a bottle cap.

A modulated, infrared, transmission photoeye system with one-shot output logic (to create the interrogation window) is chosen for this application. A modulated, infrared, convergent beam photoeye with simple on-off logic is chosen to look for the bottle cap. The transmission photoeye is placed in the light-operate mode and the convergent beam photoeye is placed in the dark-operate mode. The normally open contacts of the two photoeye output relays are connected in series to form an AND function. If both output relays are energized at the same time, a rejection cycle will be initiated.

How it Works. The sequential operation of a normal cycle (bottle cap in place) is as follows: The two photoeye systems are positioned so the convergent beam photoeye will see the bottle cap before the transmission eye will see the leading edge of the bottle. As the bottle moves down the conveyor, the convergent beam photoeye sees the cap causing its output to go low (output relay de-energizes). Next, the transmission eye sees the leading edge of the bottle causing its output to go high (relay energizes). The one-shot logic of the transmission photoeye is adjusted to cause the output to go low again (relay de-energizes) before the bottle has traveled a distance equal to the width of the bottle cap. In other words, the convergent beam photoeye is still seeing the bottle cap. Finally, the convergent beam photoeye returns to its normal high condition (output relay energized) as the bottle continues down the conveyor. During this normal cycle, the two output relays are never energized at the same time.

If the bottle cap had been missing, and therefore not detected by the convergent beam photoeye, its output relay would have remained energized while the transmission photoeye output relay energized at the leading edge of the bottle. With both output relays energized at the same time, a rejection cycle would have been initiated.

NOTE: Photoelectronic manufacturers provide a wide variety of general purpose logic output modules. These modules are solid-state and much faster than electromechanical relay outputs. For high-speed interrogation applications, you may wish to consider using faster output devices.

Similar Applications. This system can be used for missing parts detection, level height detection, and other sensing applications requiring an interrogation window.

Double Carton Detection

On a carton folding/gluing line, a method is needed to detect if two cartons (to be glued) have stuck together before the folding/gluing machine makes an attempt to fold them. A double carton will cause a machine jam. The cartons to be folded are only 20 to 50 mills thick, so conventional electromechanical caliper switches will not sense the differential. Also, the carton speed can approach 80,000 per hour, so electromechanical devices would be too slow.

A high power, modulated, infrared, transmission photoeye with a solid-state, on-delay output is chosen for this application. The transmission photoeye has a range of approximately 120 feet. It is mounted to look through the individual cartons at the beginning of their travel down the gluing line. The on-delay, solid-state output of the photoeye is connected in series with the emergency stop momentary pushbutton for the gluing line. The on-delay time period is adjusted for approximately .5 second (just long enough to activate the emergency stop). The photoeye is placed in the light-operate mode.

How it Works. The sensitivity of the transmission photoeye is adjusted high enough to burn through a single carton mass, but not high enough to burn through two cartons. (The infrared radiation of modern photoeyes can penetrate, or burn through, folded cereal boxes, single layers of heavy cardboard, etc.) As long as single cartons pass down the gluing line, the photoeye will see through them and keep its output energized. If a double carton attempts to pass, the light will be blocked causing the output to de-energize. The on-delay action is necessary for this application because of the speed the cartons travel. At 80,000 cartons per hour, a double carton could completely pass the photoeye before the emergency stop action has time to initiate. (In most machines, pressing the emergency stop pushbutton drops out a latched relay. It takes approximately 20 to 60 milliseconds for an electromechanical relay to drop out.) To ensure the photoeye output stays de-energized long enough to initiate an emergency stop, the on-delay action keeps the output de-energized for .5 second even if the carton has already passed the photoeye. The gluing machine quickly stops for removal of the double carton before it arrives at the folding section.

Since the photoeye output automatically returns to its normally energized state, the operator need only press the start pushbutton after the double carton is removed.

Similar Applications. This method can be used for detecting the proper fill level in sealed cereal boxes or similar containers.

Stacker Height Control

Flat, printed sheets are rapidly being stacked on a pallet. As the stack gets higher, a method is needed to automatically lower the stacker table.

A modulated, infrared, transmission photoeye system with on-delay output logic is chosen for this application. The photoeye is simply mounted to look above the stack at the maximum desired height. The photoeye output causes the stacker table to lower as long as it is energized.

How it Works. As the printed sheets are falling on the stack, the photoeye will be momentarily blocked, but the on-delay logic keeps the output from energizing until the photoeye is continually blocked. The photoeye will not be continually blocked until the height of the stack reaches the height of the photoeye. When this occurs, the photoeye output energizes and starts lowering the stacker table. When the stacker table is low enough for the photoeye to see above it again, the output de-energizes and stops lowering the stacker table.

Similar Applications. This system can be used for similar automatic height and level control.

Sheet Break Detection

Transmission and retroreflective type photoeye systems are used extensively in continuous sheet processes to detect the presence of the web. If a sheet break should occur, the machine is automatically stopped or an appropriate alarm is given to the operator.

Automatic Carton Gluing System

When cartons are glued and folded, the glue must be applied precisely on the glue flaps. A method is needed to control the start time and duration time of the glue ejectors.

A modulated, infrared, transmission photoeye system with solid-state latched output logic is chosen for this application. An incremental encoder and two high speed preset counters (#1 and #2) are also incorporated. The latched photoeye output is used to enable both preset counters to see the pulses being generated by the incremental encoder. The #1 preset count starts the glue ejector and the #2 preset count stops the glue ejector and resets the latched photoeye output.

How it Works. When the photoeye sees the leading edge of a carton moving down the gluing line, its output latches on. This output enables the encoder pulses to be applied to the inputs of both preset counters. The encoder is mechanically connected to the speed drive for the gluing line, so each output pulse is directly proportional to an exact length of travel by the carton to be glued. Therefore, the #1 preset counter can be preset to a number which will be equivalent to the exact spot desired to start the glue ejector. (Since you are measuring distance rather than speed, the speed of the gluing line will have no effect on this start position.) The #2 preset counter is preset higher than the #1 counter and its output will stop the glue ejector. In other words, the #2 counter is controlling the duration time. The output of the #2 preset counter also resets the latched photoeye output. With the photoeye output reset, the system is now ready for the next carton.

Similar Applications. This system can also be used in automatic taping operations.

Box Sorting

A retroreflective photoeye can be placed to see reflective (aluminum) tape on the side of boxes to be sorted. The reflective tape can be applied in vertical strips to form an on-off code. The output of the photoeye will output this code (or pattern) for automated sorting operations.

Wrong Carton Detection

Occasionally, a similar (but wrong) carton may be mixed up with the desired cartons to be packaged. A method is needed to automatically detect any wrong cartons and either inform the operator or stop the packaging line. All cartons are coded with a printed positional black mark on the glue flap. (This is commonly referred to as color mark registration.) For example, one type of carton may have the black mark 1 inch from the front edge while another carton may have the mark 2 inches from the front edge, etc.

The logic incorporated in this application is identical to the logic used for the Bottle Cap Present-application previously discussed. The primary difference is the method used for seeing the color mark registration with the convergent beam photoeye.

For color mark registration, you should use a visible convergent beam photoeye. (Infrared photoeyes cannot detect color differences.) It is also important to mount the photoeye at a slight angle (about 15 degrees) in both axis from perpendicular to the surface of the viewed object. The best angle for mounting may vary depending on the manufacturer and the type of carton. Therefore, you will have to experiment using various angles.

In this application, both angles are approximately 15 degrees. The convergent beam photoeye is mounted on a sliding adjustment rail which maintains the correct angles while allowing the operator to slide the photoeye forward or backward for set-up on various carton types.

The interrogation window is created by a modulated, infrared, transmission photoeye which is mounted to see the leading edge of the carton. A very short duration one-shot pulse is output from this photoeye when the leading edge of the carton is detected.

How it Works. The operator sets up the system for a particular carton by positioning the convergent beam photoeye so it sees the registration mark at the exact time the leading edge of the carton is detected by the transmission photoeye. If the outputs of both photoeyes do not occur at the same time, the registration mark is in the wrong position indicating a wrong carton.

Similar Applications. This system can be used for any color-mark registration application.

In the following section, applications incorporating inductive proximity sensors will be examined.

Valve Position (Open or Closed)

A method is needed to inform an operator in a remote location if a valve condition is open or closed. The valve stem is operated by a dc motor and gear reducer. It takes approximately 20 full rotations of the valve stem to completely open or close the valve. The valve stem is threaded causing it to move up or down as the valve position is changed.

Two inductive proximity switches are chosen for this application. A metal collar is mechanically attached to the valve stem. This collar moves up or down with the valve stem. One proximity switch is mounted at a position to sense the metal collar when the valve is in the closed position. The other proximity switch is mounted to sense the collar when the valve is in the open position.

With single turn valve applications, a metal cam can be fabricated for attachment to the valve stem. The two proximity sensors sense the high spot on the cam and indicate the condition of the valve accordingly.

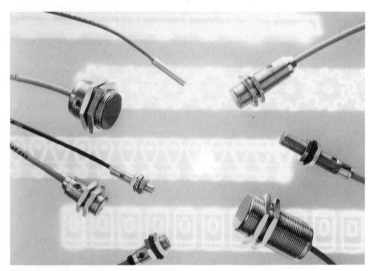

Examples of the wide variety of tubular proximity sensors (courtesy of Eaton Corporation, Cutler-Hammer Products).

Similar Applications. This method can be used for any type of position sensing of ferrous metallic objects.

Zero Speed Switch (Minimum Speed Switch)

On a continuous punch machine, a method is needed to inform the operator when the rotating roll shaft is malfunctioning.

An inductive proximity switch with one-shot output logic is chosen for this application. A 32-tooth gear is mechanically attached to the rotating roll shaft. The proximity switch senses the teeth of the gear as it rotates giving 32 outputs for each rotation of the roll. The rotational speed of the roll is adjusted for the minimum desired speed for machine operation. The one-shot output of the proximity switch is then adjusted for a minimum time period to maintain a constant output. In other words, the one-shot is continually being re-triggered before it has time to time out. The normally closed output contacts of the one-shot is used to initiate the operator alarm.

How it Works. If the rotational speed of the roll drops below the minimum acceptable speed, the one-shot output logic will have time to drop out before it is re-triggered. Since the operator alarm is initiated from the normally closed contacts, the alarm will begin to flash if the roll drops below speed. If the roll stops, the alarm will be on continuously.

Similar Applications. By utilizing the normally open contacts from the proximity one-shot output (and possibly adjusting for a shorter one-shot time period), an overspeed indication can be obtained.

For some counting applications or low-precision positioning applications, this method of creating pulses from a proximity switch sensing gear teeth can be used instead of an incremental encoder.

Thread Break Detection

On a garment weaving machine, many threads are being input simultaneously. If only one thread breaks, the finished garment will be ruined. A method is needed to automatically stop the machine if a thread breaks.

An inductive proximity switch with simple on-off logic is chosen for this application. A small hinged rod with a spool on the end of it is allowed to ride on each thread. The spool rotates with the thread and the rod is held on top of the thread by gravity. (This assembly is commonly called a *dancer arm*.) If one of the threads break, the dancer arm will fall down onto a secondary bar. This secondary bar is easily tripped and will consequently fall. The proximity sensor is mounted to sense the secondary bar when it trips. The output of the proximity sensor is used to stop the machine.

Similar Applications. This method can be used for any type of strand-breakage detection.

Cam Sensing

In any cam sensing application, proximity sensing is preferred over photoelectronic methods due to the insensitivity to contamination inherent with proximity sensors.

NOTE: If photoelectronics are incorporated in a cam sensing application, the preferred method is to use a transmission type photoeye. The photoeye is mounted perpendicular to the cam in such a manner that the beam is trying to look through the side of the cam. The rotation of the cam will consequently break the beam depending on how the cam is positioned.

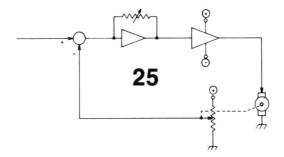

25

DC Motor-Speed Control

BEFORE BEGINNING TO EXAMINE THE BASIC TYPES OF INDUS-
trial motor-speed control systems, you should be aware of the
basic concepts and terms used with any variable-speed motor application. These basic concepts apply to both dc and variable-
frequency type motor controls.

TORQUE

Torque is the rotational force existing at the drive shaft of a
motor during motor operation. It is the turning force a motor shows
when power is applied to the motor. This turning force (torque)
is measured in units called foot-pounds (ft-lbs). A simple way of
understanding foot-pounds is to examine Fig. 25-1. Here, a one
foot long rod has been attached perpendicularly to the drive shaft
of a motor. The other end of this rod is attached to a 1 pound weight.
If the motor shaft begins to turn in the proper direction to oppose
the force of gravity on the one pound weight (in other words, lift-
ing the weight up), the motor must produce 1 foot-pound of torque
to do so. Simply stated, you should assume the 1 foot long rod connecting the motor drive shaft to the 1 pound weight does not possess any weight itself. See Fig. 25-1.

If the horsepower and base speed of a motor is known (these
parameters can usually be obtained from the motor nameplate), the
torque of the motor can be easily calculated using the following
formula:

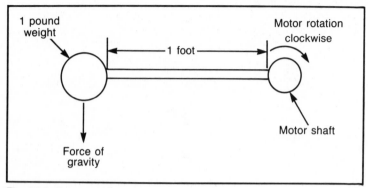

Fig. 25-1. Demonstration of torque.

$$\text{Torque} = \frac{\text{Horsepower} \times 5250}{\text{rpm (base speed)}}$$

When you consider the intended application for a variable-speed, motor-control system, you should begin by determining some basics involving the nature of the application. For example, the application could require a constant torque, constant speed, variable torque, or constant horsepower motor-control system.

CONSTANT TORQUE APPLICATIONS

A *constant torque application* is one which requires a motor to produce a constant rotational torque regardless of speed. For example, if a torque of 10 foot-pounds is required at a rotational speed of 100 rpm, the same 10 foot-pounds of force will be required at 500 rpm. A good example of a practical constant torque application is the operation of a positive displacement pump or compressor. Positive displacement pumps and compressors present the same amount of "drag" regardless of the speed at which they are being operated. Therefore, the same amount of torque is needed to overcome the drag throughout the entire operational speed bandwidth.

CONSTANT SPEED APPLICATIONS

Constant speed applications are the most common motor control applications in industry. A *constant speed application* is simply an application requiring a motor control system to maintain a constant rotational speed (measured in rotations per minute or "rpm") regardless of changes in torque requirements. Put another way, the

290

device being driven by a motor is usually referred to as the "load." In a constant speed application, if the torque requirement increases, it is often said the motor's load increases. Likewise, if the torque requirement decreases, the motor's load decreases. In either case, it is the function of a constant-speed motor control to maintain a constant speed regardless of load variations.

VARIABLE TORQUE APPLICATIONS

Variable torque applications are also very common in industry. In this type of application, the torque requirements from the load will change with changes in rotational speed. For example, consider a simple exhaust fan. At very low rotational speeds, a typical exhaust fan presents very little drag because it "sees" very little air resistance. The torque requirement in this case would be minimal. But as the rotational speed of the exhaust fan increases, its drag also increases causing a proportional increase in the torque requirement to maintain its rotation at the higher speed. Therefore in this application, the torque requirements will vary by extreme amounts depending on the rotational speed.

CONSTANT HORSEPOWER APPLICATIONS

Constant horsepower applications are very rare in industry today. However, there are a few of them left then basic concept is worth discussing.

Special motors and motor-control systems were developed which allowed a constant transfer of rotational horsepower to a load regardless of the load variations. This created some unusual (and sometimes desirable) effects. For example, if you connected a constant horsepower system to a variable load, the rotational speed of the motor would vary "inversely proportional" to the load variations. The transfer of power from the motor to the load would remain constant. In other words, if the load increased, the motor would slow down to maintain the same power transfer to the load. Consider the following practical example. If a motor is driving a load and the load increases (presents more drag to the motor), the motor must transfer more energy (horsepower) to the load to maintain the same speed. A constant horsepower system will only supply a constant horsepower to a load. Therefore, the rotational speed of the load will decrease proportionally until a new rotational speed is established. This new rotational speed represents the same horsepower utilization as was previously used before the load increased.

You can prove this by performing a transposition of the torque formula:

$$\text{(rpm) (Torque)} = \text{Horsepower} \times 5250$$

Since the horsepower is constant and the number 5250 is a constant, the right side of the equation will not change for a given application. Therefore, looking at the left side of the equation, if the torque requirement increases, the motor speed (rpm) must decrease to maintain the equation's equality.

The most common type of constant horsepower system utilized a special type of ac motor called a "Wound Rotor Motor." These systems were variable speed, but at any given speed setting the wound rotor motor operated in the constant horsepower mode.

A dc motor-control system is presently the most common means of providing variable speed for industrial processes. Dc motors are efficient, reliable, versatile, and affordable. For new applications requiring extremely wide speed variations and high torque throughout the entire speed band, a dc motor-control system is still the best answer.

HOW IT WORKS

The basic operating principle of a typical dc motor-control system is as follows. The incoming ac voltage (single or three-phase) is applied to a bridge rectifier assembly consisting of a combination of diodes and SCRs. By varying the phase angle of the firing pulses to the SCR gates, the dc output of this bridge can be controlled from zero output to the full wave rectified output of the incoming ac.

DC MOTOR CONSIDERATIONS

There are three basic types of dc motors: *series*, *parallel*, and *compound*. A series-type dc motor has its field winding in series with the armature winding. This type of motor is found in automobile starters and industrial forklifts. Its primary advantage is extremely high torque. The main disadvantage of a series motor is the poor speed regulation in variable load applications. (Under a no-load condition, a series motor will continue to increase speed until it destroys itself.) Because of these disadvantages, the uses of series motors in industry are very limited.

By far, the most common type of dc motor found in industry

is the parallel motor (also called *shunt wound*). A parallel motor has a different current flow through the field winding than through the armature. Typically, the field current is held constant and is much smaller than the armature current. (A permanent-magnet dc motor is operationally the same as a parallel dc motor with the exception it does not require a field winding. The permanent magnet establishes a permanent magnetic field instead of an electromagnetic field created by a field winding.) The speed of a parallel motor is primarily a linear relationship to the armature voltage. If the voltage to the motor armature is held constant, large load variations will have little effect on the rotational speed of the motor (typically 1 percent to 5 percent speed variation from no-load to full-load conditions). In applications requiring controlled speed with large load variations, a parallel dc motor is an excellent choice.

Another important advantage of the parallel motor is the wide variation of controlled rotational speed possible. A parallel motor is continuously speed variable from zero speed to full-rated speed. This could be an important advantage over a variable-frequency drive system which cannot be operated at extremely low speeds.

The third dc motor type is the compound motor. A compound motor is actually a combination of the series and parallel motors. A small series winding is incorporated to improve the low speed torque characteristics while maintaining a large parallel winding to take advantage of the voltage controlled speed characteristics of the parallel motor.

During normal operation of a dc motor, some arcing around the commutator is normal. For this reason, a dc motor cannot be used in an intrinsically safe environment where it could present a potential explosion hazard.

DC MOTOR MAINTENANCE CONSIDERATIONS

You must properly maintain dc motors to provide optimum results. The most vulnerable area of a dc motor is the commutator. The *commutator* consists of a group of fingers (electrical contact plates) which provide electrical contact between the brushes and the armature windings. You must check the commutator periodically and keep it clean. If the commutator becomes badly scratched or damaged (usually a result of excessive brush wear), you must replace the commutator or have it turned down on a lathe by a qualified motor-repair facility. The commutator can also develop a short if conductive contamination builds up between the commutator fingers. If this occurs, the resultant arcing across the fingers will probably destroy the commutator.

Another critical preventive maintenance routine is to periodically examine the condition of the brushes. Since all the armature current must pass through the brushes, it is very important to ensure a good solid contact between the brushes and the commutator. A poor contact will be evident by excessive arcing between the brushes and the commutator during normal motor operation.

It is normal for the brushes (constructed of soft carbon material) to wear. Dc motor manufacturers specify a minimum brush length for proper operation. During periodic brush examinations, you should check the length and any brushes measuring shorter should be replaced. (Allowing brushes to wear too short will damage the commutator.)

DC MOTOR CONTROL

By varying the field voltage or the armature voltage (or a combination of both), the speed of a parallel or compound wound dc motor can be controlled. During the remainder of this section, only parallel or compound type dc motor-control systems will be discussed.

Field voltage speed control is usually not desirable due to the nonlinear speed characteristics which result. There is also a greater chance of destroying the motor during abnormal operating conditions. (Obviously, field voltage speed control is not possible with a permanent magnet motor.)

Armature voltage speed control is by far the most commonly used dc motor-control system. In most cases, speed regulation of 1 percent to 2 percent can be obtained by simply maintaining a constant armature voltage (with IR compensation). Also, the speed to voltage relationship is highly linear. The primary disadvantage of this system is the high current requirement relative to the field.

A TYPICAL DC MOTOR-SPEED CONTROL

A dc motor control is very much like a high-power light dimmer control. This is shown in Fig. 25-2. The incoming ac power provides the high power needed for the armature through the SCR/diode rectifier bridge circuit. It is also used as a reference for synchronization of the SCR gate firing circuits. The same ac power will also be rectified and used to supply the constant dc field voltage.

The customer speed control block is essentially a speed poten-

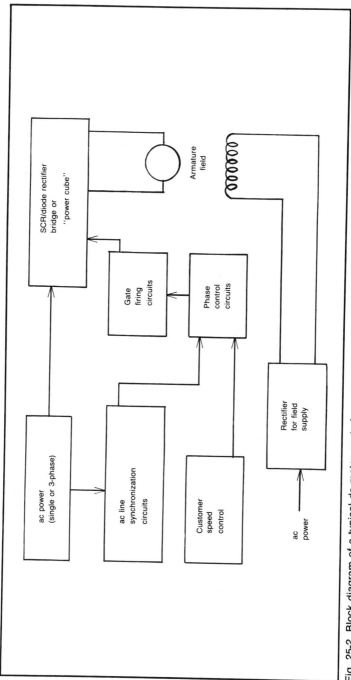

SCR/diode rectifier bridge or "power cube"

Armature field

Gate firing circuits

Phase control circuits

ac power (single or 3-phase)

ac line synchronization circuits

Customer speed control

Rectifier for field supply

ac power

Fig. 25-2. Block diagram of a typical dc motor control.

tiometer which the operator adjusts to obtain the desired speed. This signal, together with the ac line synchronization circuits output, is applied to the phase control circuits block. The output of the phase control circuits block is in the form of pulses. These pulses are phase controlled and steered to the appropriate SCR so that the output of the SCR/diode block is the correct dc voltage to provide the desired rotational speed from the dc motor.

The gate firing circuits block is usually made up of pulse transformers to provide isolation and voltage gain. The secondaries of these transformers are connected to the gates and cathodes of the SCRs in the SCR/diode rectifier bridge block.

Some modern dc motor controls use a power cube instead of discrete diodes and SCRs in the high power controlled rectifier stage. A *power cube* is a hybrid device which operates exactly the same as the discrete SCR/diode circuit. The main advantage of a power cube is lower cost.

In Fig. 25-3, a more detailed diagram of a dc motor control is shown. The incoming ac power is applied to the SCR/diode bridge rectifier circuit. By varying the firing phase angle, in reference to the ac line, the dc output of the bridge can be varied from zero volts to the maximum rectified line voltage.

The incoming ac power is also applied to isolation transformer T1. Depending on the ac phase of the primary, the secondary steering diodes (D4 and D5) will only allow current to flow through one pulse transformer (T2 or T3) when Q1 fires. The UJT (Q1) will only conduct when the voltage across C1 or C2 reaches its firing point (V peak). The speed at which this firing voltage is reached is dependent upon the charge rate controlled by the variable charge current supply block. This charge rate is what varies the phase angle of the firing pulses applied to the SCRs in the high-power bridge circuit.

In simpler terms, the steering diodes D4 and D5 determine which SCR will fire depending on whether the ac line is in the positive or negative half cycle. The charge rate of C1 and C2 control the point (or phase angle) during the half-cycle when firing actually occurs.

The output of the variable charge current supply block is controlled by the operator speed potentiometer. This potentiometer is mounted close to the remainder of the machine controls for easy access by the operator. In addition to the speed potentiometer, most motor controls also provide a minimum and maximum speed control potentiometer. You adjust these controls to set any speed limits

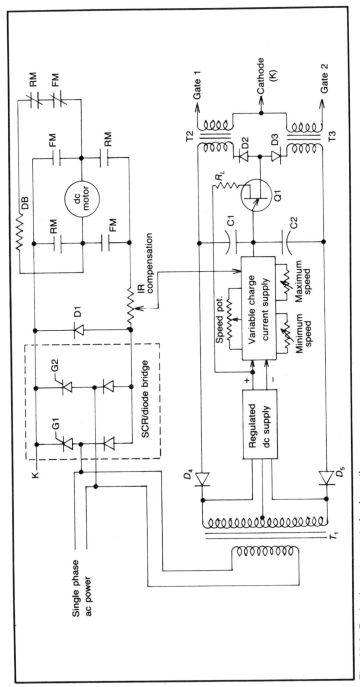

Fig. 25-3. Basic dc motor-control schematic.

297

desired. They are meant for a one-time adjustment and are usually located inside the motor control enclosure. In some cases, the maximum speed adjustment will be called the *maximum volts adjustment.*

Although the dc resistance of the armature windings is extremely low, it may have an effect on the speed regulation when the motor is heavily loaded. For this reason, virtually all dc motor controls will incorporate an IR compensation adjustment, as shown in Fig. 25-3. The IR compensation control monitors the armature current being supplied to the motor. This current signal is then used to subtract a percentage of the armature feedback voltage to compensate for the resistive voltage drop of the armature. You should adjust the IR compensation while the motor is operating under load. If it is adjusted too low, poor regulation will result. If the IR compensation is adjusted too high, the motor will oscillate.

The contacts associated with the motor armature (fm and rm), shown in Fig. 25-3, are the forward and reversing contacts. These contacts provide the same function as physically reversing the armature leads. (Changing the polarity of the voltage applied to the armature or field in a dc motor will change the rotational direction.)

In most dc motor-control applications, it is important to stop the motor rotation as quickly as possible. A simple means of accomplishing this is called *dynamic braking.* The dynamic brake resistor (db in Fig. 25-3) is connected across the motor armature by the normally closed contacts of rm and fm. Whenever rm and fm are both de-energized (meaning you are not trying to run the motor), the motor armature is disconnected from the motor control and resistor db is placed in parallel with the motor armature. Since the field voltage is still present, the motor becomes a dc generator. The resistor db presents a large load to the dc generator causing the resultant drag to be very high thus bringing the motor to a quick stop.

NOTE: The drag, or opposition to rotation presented by a generator is directly proportional to the generator load.

If a current limit control is provided, you should adjust this control for the maximum armature current desired for a particular application. Current limit control is obtained by placing a precision resistor (small value) in series with the armature. The voltage developed across this precision resistor will be proportional to the armature current drawn by the motor. This voltage is simply compared with a preset value (current limit adjustment), and if it exceeds this value, the armature voltage is automatically reduced.

REGENERATIVE DC MOTOR CONTROLS

Regenerative dc motor controls are capable of four-quadrant operation. This is accomplished by the incorporation of two SCR power bridge assemblies. By using two power bridges, it is possible to control current to the motor armature in either direction, as shown in Fig. 25-4. They are also capable of taking power from the motor in either direction of rotation and regenerating it back to the line.

This power exchange capability provides for full four-quadrant operation. This is shown in Fig. 25-5. A smooth transition from positive torque acceleration in one direction to negative torque deceleration is accomplished through the action of the SCR bridges (sometimes referred to as power converters) without mechanical contactors or heat dissipating, dynamic-braking resistors.

The ability to take power from the motor and regenerate it back to the line is often called *regenerative braking*. Motor controls with regenerative braking do not need a dynamic brake resistor.

DC MOTOR CONTROL VARIATIONS

Due to the popularity of dc drives in industry, you will likely encounter many variations (depending on the manufacturer) of the typical dc motor control previously described.

In smaller, fractional horsepower dc motor controls, the firing

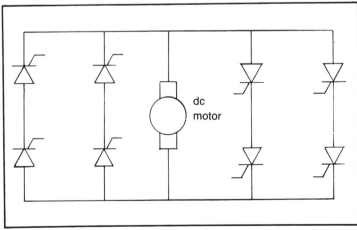

Fig. 25-4. Typical output stage of a regenerative dc motor control.

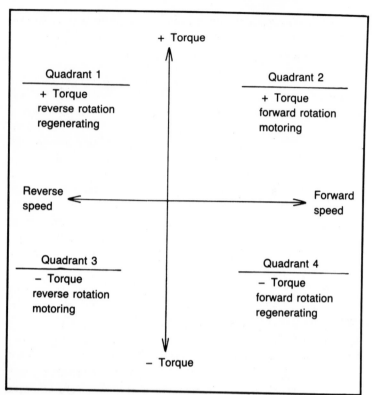

Fig. 25-5. Four quadrant operation of a regenerative dc motor control.

pulses applied to the SCRs in the high power section may not be steered according to the polarity of the incoming ac power. The gate leads of the SCRs may be simply tied together causing them to receive the gate pulses simultaneously. Since only one SCR can be forward biased at any instantaneous point in time, only the forward biased SCR will fire.

Some manufacturers provide dc motor-speed controls for very small motors which only provide controlled half-wave dc output. In these controls, only one SCR is needed for the output stage. Half-wave regenerative dc motor controls utilizing two back-to-back SCRs are also available.

EDDY-CURRENT, SPEED-CONTROL SYSTEMS

Although an eddy-current, speed-control system is not precisely classified as a dc motor control system, it is close enough in basic operating principles to be included in this chapter. This is primarily because an eddy current control is basically a small dc motor control with tachometer feedback.

A basic eddy-current speed control is illustrated in Fig. 25-6. A constant speed ac motor drives an electromagnetic clutch which in turn drives the load. The purpose of the clutch is to provide a variable "slippage" which is proportional to a dc voltage supplied by the electronic eddy current control section. The tachometer is mechanically connected to the drive shaft of the load and outputs a voltage proportional to the rotational speed of the load. The output of the tachometer is monitored by the electronic eddy current control section together with the desired speed setting provided by the operator speed control.

To help you better understand the principle of operation, assume a speed change has just been made by adjusting the operator speed control. The electronic eddy current control section compares this new speed setpoint with the tachometer output voltage and produces an error voltage. This error voltage is used to either increase or decrease the pulsating dc output voltage from a SCR/DIODE bridge in exactly the same manner as in the dc motor controls described earlier in this chapter. This pulsating dc output voltage is then applied to the eddy current clutch to either increase or decrease the slippage causing a consequential change of rotational speed to the drive shaft of the load. Since the tachometer is mechanically connected to the load's drive shaft, the tachometer voltage being monitored by the electronic eddy current control section will also change in the proper direction to reduce the error voltage. This process continues until the error voltage drops to a negligible level and the rotational speed of the load stabilizes. The biggest advantage of using an eddy current speed control is that very little power is needed to control the rotational speed of a very large load. The disadvantages of this system include poor efficiency (due to losses in the clutch) and stability when confronted with transient load variations.

DC MOTOR CONTROL INSTALLATION

When a new dc motor control is purchased and installation is required, you should closely follow the manufacturer's installation instructions.

Although solid-state dc motor controls are usually very reliable, a small installation error can destroy a large portion of the

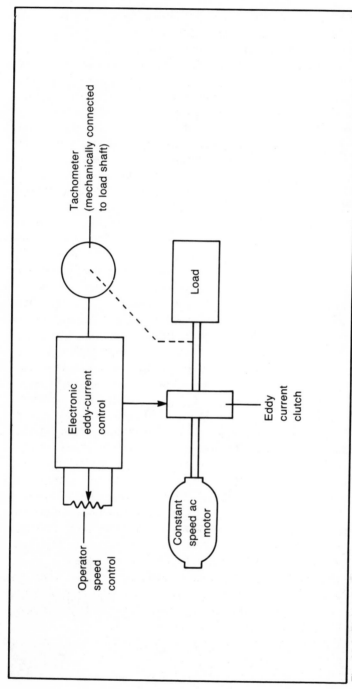

Fig. 25-6. Block diagram of eddy current speed control.

control in a matter of a few milliseconds. There are no second chances once the line power is applied to the control. For this reason, you should check and double-check to be certain all the manufacturer recommended installation procedures have been followed. If you encounter any problems or have any questions during the installation phase, you should consult the manufacturer before you apply any system power.

TROUBLESHOOTING DC MOTOR CONTROLS

NOTE: If you, as a qualified maintenance technician, encounter operational difficulties with a dc motor control, there are numerous troubleshooting methods available. This section describes some of these methods. *This section is not applicable if the problem exists on a new installation.* The author assumes the motor control with the operational difficulty has been installed and has operated properly for a reasonable period of time.

The first step in isolating a dc motor control problem is to be sure that the problem is actually in the motor control itself. A common problem found with dc motors is the brush/commutator assembly will open. This can happen if you allow the brushes to wear too much before you replace them. This problem can also occur if the commutator becomes too contaminated to conduct properly. If, in actuality, the commutator/brush assembly is open, many dc motor controls will not show any dc voltage output. This is because the SCRs will not have any current path to conduct through. If SCRs cannot conduct a current above their minimum holding current specification, they will never turn on and the dc motor control will not show any output regardless of the speed-control setting.

A simple test used to verify continuity through the commutator and armature is as follows. *Turn off all power to the control. Verify the power is actually off by measuring the ac input power to the control.* When you are satisfied that all power to the control is off, check the resistance through the motor armature leads. The dc resistance of the armature should be very low (from several ohms down to almost zero). If you obtain this reading, the brush/commutator assembly is probably in good working condition.

While checking for armature continuity, it is also a good idea to visibly check the condition of the commutator. The commutator should appear shiny to dull-gray and relatively clean. If a circular groove has formed around the commutator where the brushes are making contact, the commutator needs to be turned down by a qualified electric motor repairman. The brushes should be making good solid contact with the commutator fingers. If possible, rotate the motor shaft and be sure all the commutator fingers are clean. In some cases, only one or two armature windings may have shorted. If this happens, some of the commutator fingers will darken and give the commutator a checkered appearance. If some of the commutator fingers are considerably darker than others, the motor needs to be rewound before any further troubleshooting procedures are attempted.

A final check of the motor is to verify that there is no continuity from the motor armature to the motor shaft or motor casing. An ohmmeter can be used for this purpose, but a high voltage megger is recommended to insure a high voltage breakdown is not occurring within the motor. If a megger is used, remember to disconnect the motor control entirely from the motor to prevent the high voltage from the megger from feeding back to the solid-state components within the motor control and possibly damaging them.

If the motor appears good after the previous tests, the next step is to verify the proper ac input power is being applied to the motor control. Usually, the motor control will specify the proper ac operating voltage on a nameplate attached to the control. If not, obtain this information from the operating manual supplied with the motor control or contact the manufacturer. *The ac input voltage to most dc motor controls is lethal. Use extreme caution when you measure these high voltages.*

If there has been no apparent physical damage to the motor control (burned wires, components, or PC tracks), apply power to the motor control. If any loud humming is heard or smoke is present, turn off the power immediately. If everything appears normal, use a voltmeter to measure the incoming ac voltage. (Remember, to check all three phases if it is a three-phase unit.) If no problems have been located in the motor or the incoming ac, you can be reasonably sure the problem is in the motor control.

In the following troubleshooting procedure, the fault symptoms will be described with the corresponding description of the area of probable cause.

Important: When using any line powered test equipment to perform the following tests, always use an isolation transformer to power the test equipment. If an isolation transformer is not used, severe equipment damage and possible bodily injury could result. *Always turn off all power to the motor control before disconnecting or re-connecting any devices for testing purposes.*

1. *When you apply power, nothing happens; no dc output voltage to the motor armature regardless of the speed control adjustment.* Turn off all power to the motor control. Check all motor control fuses and circuit breakers.

   If the dc motor has a field winding, measure the field winding resistance to verify continuity and compare this value with the specified value on the motor nameplate. If this check is good, apply power to the motor control and be certain the correct field voltage and field current are present.

   Most motor controls contain a magnetic contactor (usually called the M contactor) which when energized, provides electrical connection between the motor armature and the motor control. (If dynamic braking is incorporated, the M contactor will connect the db resistor to the motor armature when the M contactor is de-energized.) The M contactor coil will be in series with many of the external operator controls such as the start, stop, emergency stops, and the internal motor thermostats. If any of these external controls (or devices) in series with the M contactor coil are defective or in the wrong operating position, the M contactor will not energize and the motor control can never apply any dc voltage to the motor armature. Consequently, the next step in troubleshooting the motor control is to verify that the M contactor is energizing. If it is not energizing, isolate the external control or thermostat creating the open in the M contactor coil circuit.

   If the M contactor is energizing and the motor control incorporates forward and reversing contactors, verify that either the forward or reverse contactor is in the correct position to run the motor. In most cases, the forward and reverse contactors have some form of mechanical or electrical interlock to prevent both from operating at the same time. (This condition would short the output of the motor control.) If the forward or reverse contactor is not operating, the interlocks should be checked.

   Another common problem that can make the motor control appear dead is if an open occurs in the operator speed control circuit. The speed control is actually a potentiometer supplying a variable dc control voltage to the motor control electronics. It is this variable dc control voltage that determines the speed at which the motor will run. Check the potentiometer for internal opens or shorts and also check the potentiometer wiring. (A large number of motor control manufacturers specify a 5-kohm potentiometer for this purpose.)

2. *The motor runs at full speed and cannot be controlled by the operator speed control.* If the motor control is using tachometer feedback, check the tachometer feedback voltage. If this signal is not present, the motor control will always run at full speed. In some cases, it is easy to convert the motor control over to armature voltage feedback for speed regulation. If there is any question regarding the proper tachometer output voltage, a temporary conversion over to armature feedback may be an easy troubleshooting procedure.

In some motor controls, it is possible for the operator speed control potentiometer (or associated wiring) to short in such a way as to cause the motor control to maintain a maximum output.

3. *Motor will not run fast enough or slow enough to meet the application needs.* Check for a possible problem with the operator speed control potentiometer or associated wiring. Also check for proper adjustment of the minimum speed and maximum speed controls.

If the motor is running too slow, check the armature output voltage to verify it is lower than normal. If it is not, the motor should be checked for possible shorted armature windings. (If this was the case, the motor would probably be hotter than normal.)

If the motor is running at approximately half its normal speed, one of the components in the high power SCR/diode output bridge (or power cube) may have opened. If this condition occurs, the motor control will be providing a half-wave output instead of a full-wave output. By measuring the voltage across each component in the bridge, the open component will be the one showing an abnormally high voltage across it compared to the rest within the bridge. The same symptom would occur if one of the SCR's was not receiving the proper gate pulse to turn it on. An oscilloscope can be used to look at the gate pulses and verify their presence. *(Don't forget to use an isolation transformer to power the oscilloscope if it is line powered.)* If the proper gate pulses are not present, a problem exists in one of the prior firing circuits. If the gate pulses are present, replace the open component (or power cube).

4. *The circuit breaker intermittently trips during normal motor control operation.* An intermittent problem is, by far, the most difficult type of problem to isolate. Usually, when the main circuit breaker trips, it means a phase-to-phase (three-phase systems) or phase-to neutral (single-phase systems) short has occurred.

   In regenerative systems, this condition can be caused by firing an SCR at the wrong time (misfire) due to electrical noise within the motor control circuits. This electrical noise can be created internally by the motor control itself or generated externally by some other electrical device. Check all noise suppression devices within the motor control. Many motor controls contain a noise suppression card or snubbers for this purpose. Loose connections, especially around the gate driver circuits, can also cause this type of problem. Thoroughly check the integrity of all internal connections.

   In single quadrant drives, this type of problem is usually related to an SCR breaking down under load. (This is quite common with thyristors.) There is sophisticated equipment available to test the dynamic breakdown of thyristor devices, but it is sometimes easier and less expensive to replace them. If replacement corrects the intermittent problem, the original SCRs can be re-installed one at a time until the problem returns. It can then be determined that the last SCR re-installed was the one breaking down.

   Intermittent circuit breaker trips can also be a result of improper adjustment of the current limit potentiometer. If you suspect this condition, the current limit should be adjusted to limit the current to a lower value. Also, the obvious should not be overlooked. A few cases occur where the heaters within the circuit breaker itself become stressed over a long period of time. This causes the breaker to trip at a lower current value than originally specified. Installation of new heaters will solve the problem.

5. *The circuit breaker trips immediately upon applying ac power to the motor control.* This problem indicates a phase-to-phase (three-phase systems) or phase-to-neutral (single-phase systems) short in the high power bridge section.

This type of problem is usually caused by one of the thyristors or diodes in the power bridge developing an internal short. An ohmmeter can be used to check for shorts in this area. It is also possible for a short to develop from the armature leads to earth ground in the dc motor.

If none of the above conditions exist, it is possible one of the SCRs is immediately breaking down when power is applied (an ohmmeter may not indicate this) or an erroneous gate pulse is turning on two or more SCRs at the same time. Turn off all power to the control, disconnect all SCR gates, and disconnect one side of all of the SCRs. Reapply power to the control. If the circuit breaker does not trip, the problem can be isolated in a trial and error fashion. Turn off all power to the control, reconnect one SCR (do not reconnect any gate leads yet), and reapply power. Continue this process until all of the SCRs are reconnected or the circuit breaker trips. If the circuit breaker trips, the last SCR reconnected (or one of the last two SCRs in three-phase systems) is defective and should be replaced. (Remember, it is possible for more than one SCR to be defective. If a defective SCR is found in this procedure, continue to test the remaining SCRs.) If all of the SCRs are reconnected without tripping the circuit breaker, a problem exists in the gate pulse or phase control circuits.

The gate pulse circuits (also called firing circuits) can be very difficult to troubleshoot in three-phase and regenerative motor controls. You should contact the manufacturer for specific instructions regarding the isolation and repair of problems in this area.

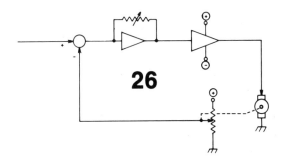

Variable-Frequency
Motor-Control Drives (VFD)

VARIABLE-FREQUENCY DRIVE (*VFD*) *SYSTEMS* ARE BECOMING increasingly popular in industry because they have some distinct advantages over more conventional dc and eddy-current clutch-drive systems. A typical VFD system is less expensive, requires less maintenance, and can provide a considerable energy savings in certain applications. This is because a dc drive will suffer from a lagging power factor at lower speeds (typically, 0.5 at half-speed and .25 at one-fourth speed). In addition, a typical induction ac motor is smaller, less expensive, requires less maintenance (no brushes or commutators), and can be used in explosive atmospheres.

SHOULD VFD SYSTEMS BE
USED FOR ALL APPLICATIONS?

No. A VFD system is not normally the optimum choice for applications requiring high torque at low speeds. At very low speeds, the frequency of the applied voltage to the motor must be severely reduced to provide the low speed. If the frequency is driven below approximately 6 hertz, many types of VFD systems may cause the motor to jerk at that frequency, causing an undesirable condition called *cogging*. Also, standard ac induction motors are inefficient at low frequencies and may overheat if maintained at low speed for long periods of time.

THE BASIC PRINCIPLE OF VFD OPERATION

The only way to appreciably change the speed of a standard ac induction motor is to change the frequency of the applied ac voltage to the motor. In other words, ac induction motors are synchronous in nature (if you consider a typical 3 percent to 5 percent slippage as negligible).

Virtually all ac induction motors manufactured for use in the United States are designed for optimum operation at the standard 60-Hz line frequency. A VFD system converts the standard 60-Hz line to a variable-frequency voltage and applies it to the ac induction motor. A typical VFD system may supply a variable frequency from a minimum of 6 Hz to a maximum of 90 Hz. This means that a standard 1800-rpm induction motor would have a speed range of approximately 180 rpm to 2700 rpm if used with a VFD control.

The torque developed in an induction motor is proportional to the concentration of the magnetic flux lines in the air gap between the stator (motor windings) and rotor (analogous to the armature in a dc motor). For maximum torque, the magnetic flux density must be as high as possible without saturating the iron material in the rotor. This flux density is directly proportional to the voltage applied to the motor windings and inversely proportional to the frequency (due to the effect of the inductive reactance of the stator). As the frequency is reduced in a VFD system, the applied voltage amplitude to the motor must also be reduced to prevent saturation. (A long-term condition of saturation would cause overheating and possibly damage the ac motor.)

For example, if a standard motor is designed to operate at 480 Vac and 60 Hz, a ratio of 480:60 or 8:1 is established. This ratio is commonly referred to as the *volts-per-hertz ratio* and must be maintained over the entire operating range of the VFD. If the applied frequency to the motor happened to be 30 Hz, the applied voltage amplitude should also be reduced by 50 percent, or to approximately 240 V.

A FUNCTIONAL BLOCK DIAGRAM OF A VFD SYSTEM

In the following section, we will examine a functional block diagram of a typical three-phase VFD system. A basic block diagram is shown in Fig. 26-1. The converter block accepts the three-phase line voltage and converts it to a variable dc voltage and current. (You could operate a standard dc motor from the output of this stage.) This is the stage responsible for providing the variable am-

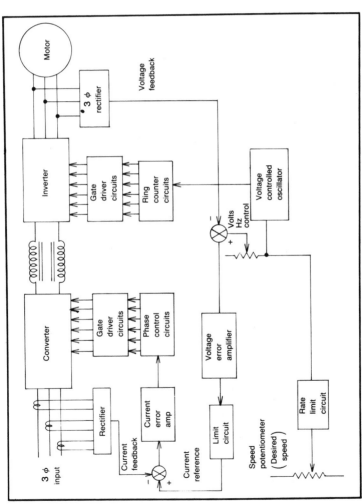

Fig. 26-1. Basic block diagram of a typical three-phase variable frequency motor control.

plitude output to the motor so the proper volts-per-hertz ratio can be maintained.

You can see how this is accomplished by examining Fig. 26-2. It should be evident that the converter stage is basically a three-phase bridge rectifier. The primary difference is that SCRs are used instead of standard diode rectifiers. By using SCRs in the converter stage, the SCR turn-on phase angle can be varied, causing the average dc output voltage to vary. In simple terms, the converter stage consists of a high-power, variable dc supply. The inductors in series with both sides of the dc bus are used to filter and average the dc output.

In Fig. 26-1, the output of the three current transformers (monitoring the three-phase ac input current) is fed into the Rectifier block. The Rectifier block converts this voltage to a proportional current feedback signal, which is compared to the current reference signal. The error signal from this comparison is fed into the Current Error Amp block. The amplified error signal is then fed to the Phase Control Circuits and Gate Driver Circuits blocks. These last two blocks control the turn-on phase angle of the SCR's in the converter stage, which proportionally varies the dc bus output applied to the Inverter stage.

In an effort to simplify this operation, you may consider the current reference as simply a setpoint. The current transformers are the sensors. The Rectifier block is the signal conditioner providing a useable feedback signal for comparison with the setpoint. The Current Error Amp block provides the proportional gain. The Phase Control Circuits and Gate Driver Circuits blocks are the actuators

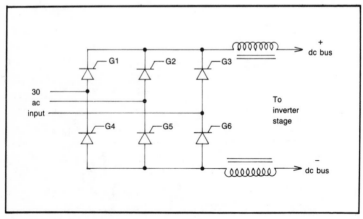

Fig. 26-3. Typical current source converter stage.

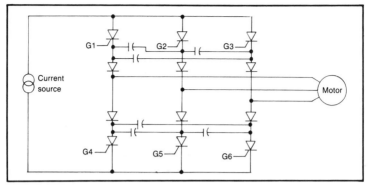

Fig. 26-2. Typical current source inverter stage.

controlling the Converter output. So, in essence, these circuits are operating as a proportional control loop.

The Inverter block accepts the dc output of the Converter stage. The Inverter block converts this dc to a variable-frequency ac output. It accomplishes this with the basic circuit shown in Fig. 26-3. This circuit also uses SCRs as solid state switches to route or steer the dc to the appropriate motor windings in the proper sequence. The purpose of the capacitors is to force the appropriate SCR to turn off when another turns on. This turning-off action is called *commutation*. Therefore, the capacitors shown in Fig. 26-3 are referred to as *commutation capacitors*. Basically, the commutation capacitors charge and discharge in such a manner as to provide a transient reverse bias to the SCRs and force them to turn off at the proper time. (SCRs cannot turn off unless the current flowing through them falls below their specified holding-current parameter.) The commutation scheme shown in this example is only one of many designs currently being manufactured.

Refer to Fig. 26-1. Examine the remaining blocks. The operator sets the speed potentiometer to whatever the desired speed. The speed potentiometer provides the desired speed signal, which is fed to the Rate Limit Circuit block. This block limits the acceleration and deceleration rates of the motor to prevent an excessive current surge through the system when a speed change is desired. The output of the Rate Limit Circuit block is a dc voltage and is fed to the Voltage Controlled Oscillator block which applies a proportional frequency to the Ring Counter Circuits block. The Ring Counter Circuits block converts the input frequency to sequential parallel output pulses in the proper phase relationship for correct turn-on of the SCRs in the Inverter stage. The Gate Driver Cir-

313

cuits block provides amplification and isolation between the Ring Counter Circuits and Inverter stages.

The only remaining function needed for proper operation of a VFD system is a means of maintaining the correct volts-per-hertz ratio. The three-Phase Rectifier block converts the ac voltage applied to the motor to a proportional dc voltage for comparison purposes. This voltage is compared with a portion of the control voltage being applied to the Voltage Controlled Oscillator block. The difference of the signal obtained is applied to the Voltage Error Amplifier block for amplification and then conditioned by the Limit Circuit block. The Limit Circuit block will also limit excessive current flow if a motor overload should occur. The output of the Limit Circuit block is the *current reference signal*. As described previously, the current reference signal is used as a setpoint to control the output of the Converter stage, which provides the correct output amplitude to the motor.

GENERAL VFD CONSIDERATIONS

The VFD system you have just examined is commonly referred to as a *current-source VFD*. In addition to this type of VFD, two other types are currently being manufactured: the *voltage-source VFD* and the *pulse-width-modulated VFD*.

VOLTAGE-SOURCE VFD

The voltage-source VFD design is essentially the same as the current-source design with one exception. Instead of using inductors in series with the converter-stage output to provide a controlled current source, the voltage source design uses a large bank of capacitors in parallel with the output of the converter stage. This provides a controlled voltage source. The inverter stage operates on the same basic principle, but the commutation method is different.

The primary advantage of a current-source VFD is that the commutation system is simpler and much less susceptible to false triggering. Its primary disadvantage is that it must be matched to the ac motor it will be controlling. In other words, a 125-hp. current-source VFD will not perform satisfactorily if used to operate a 50-hp. motor. A voltage-source VFD does not suffer from this limitation, but its communication system is much more complicated, and it is more susceptible to false triggering.

314

PULSE-WIDTH-MODULATED VFD SYSTEMS

More modern VFD systems are currently being manufactured that operate on an entirely different principle than any VFD system discussed previously in this chapter. But before you can understand their operation, you must understand the basic principles behind pulse-width modulation (PWM).

Imagine that you had a 100-watt lamp connected to a power source through an on-off switch. If you turned the lamp on for one minute, you would consume 100 watts of power during the entire minute. If you then turned the lamp off for one minute, you would consume 0 watts of power for that entire minute. If you continued to cycle the lamp in this manner (on for one minute, off for one minute) for one hour, the "average" power consumption during the hour would be 50 watts. This is logical, because the lamp was on half of the time and off half of the time.

Now suppose you varied the ratio of the on time to the off time. For example, if the on time was only 1/2 minute and the off time was 1 1/2 minutes the lamp would only consume 100 watts of power 25 percent of the time. Therefore, the average power consumption for a one-hour period would be 25 watts (25 percent of 100 watts is 25 watts).

Of course, this would not be a very desirable way of controlling power to a lighting system, because it would be flashing on and off at a visible rate. But consider using milliseconds as a cycle time base instead of minutes. For example, assume you cycled the lamp on and off at a rate of 1000 times per second, or 1 kHz. Your eyes could not respond to this rapid change, but you could still vary the average power to the lamp by varying the ratio of the on time and off time. This is the basic principle behind pulse-width modulation.

The cycle time is referred to as the *base frequency*. The ratio of the on time versus the off time is called the *duty cycle*. The base frequency does not change, but the duty cycle can vary from 0 to 100 percent. Figure 26-4 illustrates a duty cycle of approximately 10 percent (i.e., the on time represents 10 percent of the total time period). Figure 26-5 illustrates a 90 percent duty cycle. Note that the total time period is determined by the base frequency (1 kHz) and does not change in either case.

If you replaced the on-off switch to the lamp with a solid-state switch capable of switching at very high speed (assume 1 kHz), and you had an electronic control to vary the duty cycle throughout the entire range of 0 to 100 percent, you could cause the lamp to dissi-

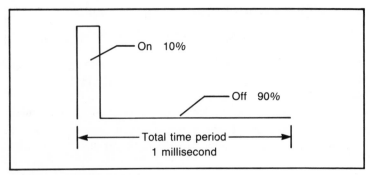

Fig. 26-4. A 10% duty cycle.

pate any average power from 0 to 100 watts. The brightness of the lamp would consequently range from total darkness to full brightness. Your eyes could not respond to the 1-kHz base frequency, and the illumination from the lamp would appear constant. In other words, you now have an expensive light dimmer.

The key word in the previous sentence is "expensive." Conventional light dimmers are much more cost-effective and operate equally well. But when it comes to controlling ac motor speeds, pulse-width modulation can provide some significant advantages over conventional thyristor-output VFDs.

Figure 26-6 illustrates a block diagram of a typical PWM variable frequency drive. As in previous VFDs you have examined, the incoming ac line power is rectified and filtered. The difference in this case is that the dc bus is not variable to compensate for the volts-per-hertz adjustment. The ac line power is simply rectified and filtered. As you should realize from your knowledge of basic electronics, the voltage amplitude on the dc bus is the approximate peak-voltage equivalent of the ac Line rms voltage level. For ex-

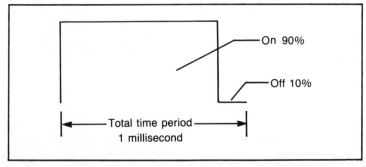

Fig. 26-5. A 90% duty cycle.

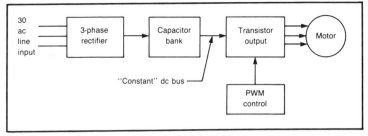

Fig. 26-6. Block diagram of a typical PWM VFD.

ample, if the ac Line voltage happened to be 440 volts, the dc-bus voltage would be approximately 622 volts dc. (440 × 1.414 = 622)

The dc bus is applied directly to the transistor-output stage. The output transistors are switched (cycled) at a rapid rate and are used to control the output duty cycle. The variations (called *modulation*) in the output duty cycle simulate a sine-wave current flow through the motor's stator windings.

Figure 26-7 illustrates an example of the PWM output voltage and the resulting current flow through the stator windings. Since the stator windings in an ac motor are inductive and are designed to operate at 60 or 50 hertz, they will tend to average the rapid cycling of the base frequency, causing the current flow to become a function of the duty-cycle modulation.

In addition to fabricating a "near" sine-wave current through the stator windings, the PWM output is also used for the volts-per-hertz compensation. This is accomplished by adjusting the duty cy-

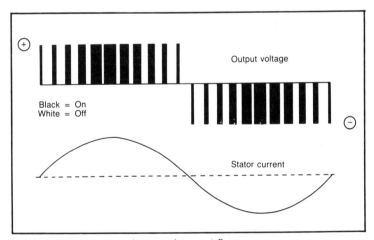

Fig. 26-7. PWM output voltage and current flow.

cle of all individual cycles by the same amount. The effect is the same as changing the output voltage level of the converter stage in a conventional thyristor VFD.

If you are confused regarding the difference between the output modulation (sine-wave current fabrication) and the volts-per-hertz adjustment, you may consider them in the following manner. Output modulation means a certain periodic pattern (a sine wave in this case) is superimposed on the PWM output voltage by the encoding of the periodic pattern into pulse widths instead of analog voltage levels. This means that each individual pulse will vary in width from the preceding pulse by some amount proportional to the level change in the periodic pattern since the preceding pulse occurred. In contrast, the volts-per-hertz adjustment is made by increasing or decreasing all of the pulse widths by an equal amount.

For example, assume a PWM VFD system is operating a 440-Vac, 60-Hz motor at a simulated 440-Vac, 60-Hz speed setting. The operator decides to reduce the motor speed by 50 percent and changes the speed potentiometer setting accordingly. The PWM VFD changes the periodic modulation rate to 30 Hz instead of 60 Hz. In addition, it must also reduce all of the pulse widths by 50 percent to simulate the desired 220-Vac volts-per-hertz level change.

As you may realize, a PWM VFD is a complex type of motor control. In addition, if the PWM VFD is designed to control a three-phase motor, the complexity becomes much greater, because of all the previously discussed functions must be controlled on three output lines, while the fabricated sine-wave stator currents must be maintained at 120 degrees out of phase from each other. For this reason, virtually all PWM VFDs incorporate a microprocessor-based controller to continuously monitor and control all motor functions.

Obviously, the biggest disadvantage of a PWM VFD in comparison to a conventional thyristor-type VFD is its complexity. On the other hand, there are some decided advantages to the PWM VFD. Conventional thyristor-type VFDs have a lagging power factor similar to typical dc drives, and their output waveshape applied to the motor is in the form of a stepping or "staircase" square wave. This square-wave output produces an abruptly varying stator-current waveshape, which can cause excessive motor heating and cogging at low motor speeds.

Figure 26-8 illustrates voltage outputs and stator-winding currents produced from a typical thyristor-type VFD. A PWM VFD

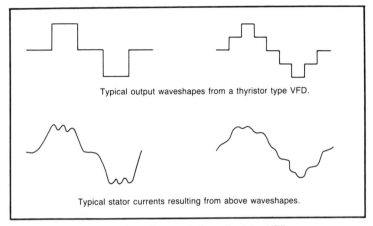

Typical output waveshapes from a thyristor type VFD.

Typical stator currents resulting from above waveshapes.

Fig. 26-8. Voltage outputs and currents from thyristor VFD.

system maintains a power factor very close to 1 at any speed, and the near-sinusoidal stator-current waveshape produces very smooth low-speed torque.

One other advantage to a PWM VFD motor control is the output power devices used. A conventional thyristor VFD uses SCRs as output devices and a commutation system to fire the SCRs at the correct time. Most PWM VFDs incorporate high-power transistors as output devices. Transistors are much easier to control than thyristors and much less susceptible to misfiring. Modern high-power transistors can be manufactured with ratings of 900 volts from collector to emitter and maximum collector currents of 200 amps.

TROUBLESHOOTING THYRISTOR-TYPE VFD SYSTEMS

In general, troubleshooting VFD systems is considerably more difficult than troubleshooting dc motor-control systems. This is because VFD systems are usually more complex than dc drives. Manufacturers of VFD systems normally provide good troubleshooting documentation with their drives, which will contain specific information regarding problem analysis. Depending on the size and complexity of the VFD system, some units contain special diagnostic panels that allow you to look at critical checkpoints within the system. The associated troubleshooting manual will describe the voltages or frequencies you should measure at these checkpoints and steer you toward the defective section.

Considering the entire scope of VFD systems, there are a few generalities regarding troubleshooting which can be discussed. As in the case of dc motor controls, if the VFD system appears dead, you should suspect a problem with the control circuits (emergency stop pushbuttons, speed potentiometers, motor thermostats, etc.) or a loss of the incoming ac power to the VFD system. Check the line fuses and main circuit breakers if an ac power loss is determined.

Virtually all VFD systems incorporate high-speed solid-state fuses on the incoming ac line. If one or more of these fuses are blown, it is critical to replace them with identical replacement fuses. If you try to use a standard fuse instead, you can severely damage the VFD system.

If you replace any of the line fuses and they blow immediately after power is applied to the system, do not continue to replace the fuses. Besides the fact that solid-state fuses are very expensive, this indicates an additional problem somewhere else in the VFD system. The first step is to isolate the problem to either the converter or inverter stage. The easiest way to do this is to turn off all power to the system and disconnect the dc bus between the two stages.

Caution: Some types of VFD systems contain a large bank of charged capacitors between the converter and inverter stages. Do not attempt to disconnect the dc bus until you have verified there are no high voltages present. Once you have disconnected the dc bus, there are several steps you may wish to take, depending on the system.

If the SCR gates are connected to the gate driver circuits by means of a plug, you may wish to unplug all of the SCR gates in the converter stage. Apply the incoming ac power to the system. If a line fuse blows again, you know the problem to be one or more of the diodes or SCRs in the converter stage. If none of the fuses blow, turn off all power to the system, and replace the gate connections in the converter stage. Once again, apply the incoming ac power to the system. If one or more of the fuses blow at this point, you know the problem is being caused by a misfire to one of the SCR's in the converter stage. If none of the fuses blow, you know the problem is in the inverter section.

In some smaller VFD systems, the converter SCR gate connections may be soldered directly to the gate driver circuits. In order to save time and work, you may not want to disconnect these gate connections before you try applying power to the system (after disconnecting the dc bus). If the problem is in the inverter stage (which is most common), none of the line fuses will blow, and you can proceed directly to troubleshooting the inverter section.

After the dc bus is reconnected, the SCR gate leads in the inverter section should be disconnected and power reapplied to the VFD system. If one or more of the line fuses blow, you know one of the SCRs in the inverter section is defective. If not, the problem is in the commutation or gate-triggering systems.

TROUBLESHOOTING PWM VFD SYSTEMS

Although PWM VFD systems are more complicated than thyristor-based VFD systems, troubleshooting procedures can often be easier and quicker than in conventional systems. (Of course, this depends on the manufacturer.) Due to their complexity, PWM VFDs are usually equipped with much better diagnostic hardware and documentation.

Isolating faults to the printed circuit board level is often accomplished by the microprocessor within the unit. A simple routine is performed by the operator to cause the microprocessor to perform checks throughout the entire system and display failure codes to inform the maintenance personnel which section needs replacement. In some systems, the microprocessor automatically performs these checks without any operator intervention as soon as a failure is detected.

The most common problem with PWM VFDs is a failure in one or more of the output devices. As stated previously, these output devices are usually high-power transistors and can be tested like any other power transistor. Some PWM VFD systems utilize high-power gate turn-off (GTO) silicon control rectifiers as output devices. A GTO operates very much like a conventional SCR except it has the capability of being turned off by short pulses of reverse gate current. You can test GTOs with a very simple "bench" circuit you can build. But before attempting this, you should contact the GTO manufacturer regarding component values and expected results.

If the microprocessor/controller section of a PWM VFD fails to operate, it obviously will not be capable of performing diagnostic routines. In this case, you should check all of the low-voltage dc power supply voltages used to power the microprocessor/controller. If they are all within normal tolerances, it may be necessary to replace the microprocessor/controller card.

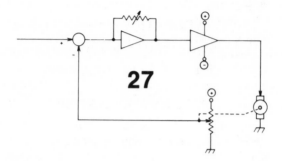

27

Calculating Motor Variables

T HE FOLLOWING SECTION PROVIDES THE READER WITH A
group of formulas which are almost essential when working
with dc or ac motors. These formulas are a troubleshooting aid as
well as a basic "rule of thumb" for new installations.

To begin, let us define a few common terms:

I = amperes E = volts   kW = kilowatts
E_{eff} = efficiency (if unknown, assume .85)
P_f = power factor (if unknown, assume .85)
hp = horsepower   kVA = kilovolt-amperes

The following section is applicable to dc motors only. If you
know horsepower and you want to calculate amperes:

$$\frac{hp \times 746}{E \times E_{eff}} = \text{amperes}$$

If you know kilowatts and you want to calculate amperes:

$$\frac{kW \times 1000}{E} = \text{amperes}$$

If you want to calculate horsepower:

$$\frac{I \times E \times E_{eff}}{746} = \text{horsepower}$$

The following section is applicable to ac motors only. If you know horsepower and you want to calculate amperes:

(single-phase motors) $\dfrac{hp \times 746}{E \times E_{eff} \times \text{pf}} = \text{amperes}$

(three-phase motors) $\dfrac{hp \times 746}{1.73 \times E \times E_{eff} \times \text{pf}} = \text{amperes}$

If you know kilowatts and you want to calculate amperes:

(single-phase motors) $\dfrac{\text{kW} \times 1000}{E \times \text{pf}} = \text{amperes}$

(three-phase motors) $\dfrac{\text{kW} \times 1000}{1.73 \times E \times \text{pf}} = \text{amperes}$

If you know kVA and you want to calculate amperes:

(single-phase motors) $\dfrac{\text{kVA} \times 1000}{E} = \text{amperes}$

(three-phase motors) $\dfrac{\text{kVA} \times 1000}{1.73 \times E} = \text{amperes}$

If you want to calculate horsepower:

(single-phase) $\dfrac{I \times E \times E_{eff} \times \text{pf}}{746} = \text{horsepower}$

(three-phase) $\dfrac{I \times E \times 1.73 \times E_{eff} \times \text{pf}}{746} = \text{horsepower}$

NOTE: To convert kVA to kW, multiply kVA by the power factor.

General "rules of thumb"

1. Although one horsepower = 746 watts, one horsepower will draw approximately one kVA.
2. A 460-volt, three-phase motor draws approximately 1.25 amperes per horsepower. At 230 volts, motors draw approximately 2.5 amperes per horsepower.
3. At 120 volts (single-phase), one kVA is approximately equal to 8.5 amperes.
4. The starting and maximum running torque of ac inductions motors will vary as the square of the voltage.
5. The speed of the ac induction motor will vary directly with the frequency.

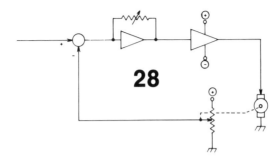

A Basic DC-Motor Servo System

VARIOUS DC-MOTOR-CONTROLLED SERVO SYSTEMS ARE widely used in industry to provide highly accurate and repeatable automatic positioning systems. The advantages associated with dc servo systems are simplicity, accuracy, and relatively low cost.

BASIC DESCRIPTION

A basic block diagram of a typical dc servo system is shown in Fig. 28-1. As shown, the setpoint signal is compared with the actual position feedback. This comparison is performed by inverting the position feedback signal and algebraically adding the two signals. If, for example, the two signals happened to be of the same amplitude, the algebraic sum would be zero. (They would cancel each other out.)

As you may have guessed, this is a perfect application for an operational amplifier. The setpoint signal could be applied to the noninverting input and the feedback signal to the inverting input. The feedback signal would be inverted and algebraically added to the setpoint signal. The output of the operational amplifier would be the difference between the two, or *error*. The gain of the operational amplifier could be adjusted to provide the desired *error signal gain*, or *proportional band*. (In older servo systems, a differential amplifier was used in place of an operational amplifier.)

The error signal may be of either polarity. If the feedback sig-

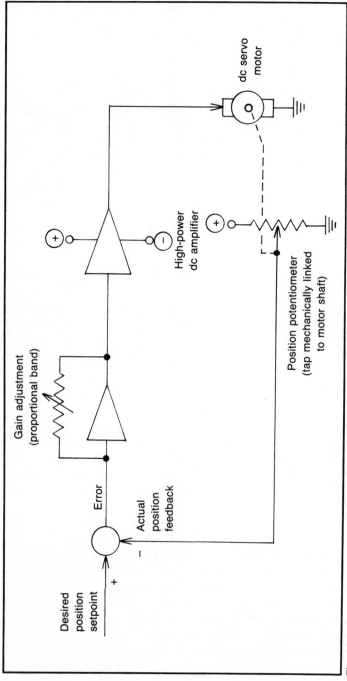

Fig. 28-1. Basic dc-motor servo system.

326

nal is larger than the setpoint signal, the output of the operational amplifier is negative. If the setpoint signal is larger than the feedback signal, the output is positive. The output of the operational amplifier (error amplifier) is applied to a high-power amplifier to provide sufficient power to run the dc servo motor.

The shaft of the dc servo motor is mechanically linked in some manner to a position potentiometer (or feedback potentiometer). As the motor shaft rotates, it adjusts the potentiometer to reduce the error signal. The direction of the motor rotation will depend on the polarity of the error signal. The motor does not have to be mechanically linked in a 1:1 ratio with the position potentiometer. In other words, depending on the application, the dc servo motor may be gear-reduced so that a single turn of the position potentiometer will require many turns of the motor shaft.

A servo system is actually nothing more than a proportional control loop. The gain of the error amplifier stage determines the proportional band. Depending on the application, integral or derivative control actions may also be incorporated.

Servo systems can be incorporated in virtually every type of automatic positioning application. If the servo system is to be controlled by a computer, the position potentiometer can be replaced with an absolute encoder to provide a parallel digital output for position feedback. The comparison between the position feedback and the position setpoint can then be performed in the computer. The error-signal output from the computer will be a digital word and must be converted to analog (with a D/A converter) before applying it to the error amplifier.

Servo systems using large servo motors will frequently incorporate the same type of adjustments as seen in standard dc motor controls. These adjustments will include IR compensation, minimum and maximum speed control, current limit, etc.

TROUBLESHOOTING SERVO SYSTEMS

You can troubleshoot a servo system just as you would troubleshoot any other type of proportional system. Verify that the gain (or proportional band) adjustment is correct. If the gain is set too high, the servo motor will oscillate. If it is set too low, the servo motor response will be sluggish or nonexistent.

The existence of an appropriate error signal is usually the key in proportional control systems. If a large differential exists between the setpoint and feedback signals without a large error signal, the problem is in the comparison stage (usually an operational amplifier or differential amplifier).

In some cases, such as in proportional control valves, safety limit switches will be used to shut down the drive to the servo motor if an excessive error causes the servo motor to try to drive beyond certain physical limits. If one of these limit switches is activated, the following troubleshooting method is recommended:

1. Turn off all power to the servo system.
2. Disconnect the servo motor from the high-power amplifier stage.
3. Physically rotate the shaft of the servo motor in the correct direction to deactivate the limit switch.
4. Connect a voltmeter to monitor the error signal.
5. Reapply power to the servo system. Try to obtain a null (zero) error voltage by rotating the servo motor shaft. If a null error voltage can be obtained, the problem is in the high-power output or gain adjustment stages. If a null error voltage cannot be obtained, the problem is probably in the position potentiometer or comparison stages.

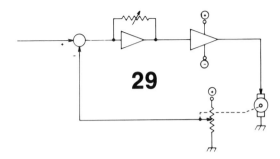

Stepping Motors

STEPPING MOTORS ARE WIDELY USED IN INDUSTRY. THE shaft of a stepping motor rotates a precise number of degrees for each drive pulse applied to it. This step, or rotational increment, is relatively independent of the drive-pulse level or minor variations in pulse width. In other words, the total number of shaft rotations is proportional to the total number of drive pulses received by the motor.

Depending on how the stepping motor is gear-reduced, each drive pulse can be directly related to a precise rotational increment. For example, a stepping motor can easily be designed to move the X-axis of a CNC milling machine exactly 1 inch for every 1000 drive pulses it receives. In this case, each drive pulse will directly relate to a 1-mill (.001 inch) movement of the X-axis. The important thing to realize in this example, is that this precise positional control is based on digital (on-off) pulses instead of analog voltage levels.

For this reason, stepping motors lend themselves very well to digital positional control applications. Before examining a stepping motor positional control system, first consider the basic operation of a stepping motor.

BASIC STEPPING MOTOR OPERATION

Refer to Fig. 29-1. Note that the armature of the stepping motor is constructed of a permanent magnetic material. The two center-tapped field windings can be electromagnetized indepen-

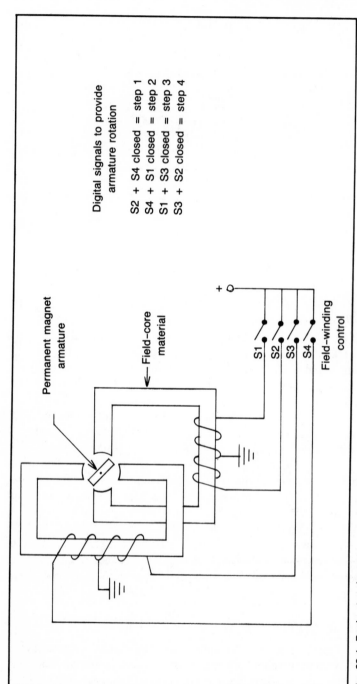

Permanent magnet armature

Field-core material

Field-winding control

S1 S2 S3 S4

+

Digital signals to provide armature rotation

S2 + S4 closed = step 1
S4 + S1 closed = step 2
S1 + S3 closed = step 3
S3 + S2 closed = step 4

Fig. 29-1. Basic stepping-motor operation.

dently in either magnetic polarity. Since there are two fields and each field can be electromagnetized in either magnetic polarity, there are four possible combinations, or steps. Two coincidental digital signals are required to generate each step. By generating these steps in the proper sequence, the electromagnetic field can be rotated around the armature.

Since the armature is constructed of a permanent magnetic material, it will follow the rotating electromagnetic field. The armature can also be made to reverse direction by changing the step sequence. In the case of the example shown in Fig. 29-1, each step equals a 90-degree rotational increment by the armature. Also, the rotational speed is directly proportional to the speed at which the motor is stepped.

Unfortunately, stepping motors have limitations. The torque of a stepping motor is inversely proportional to speed. The torque characteristics become even worse if the speed is increased to the resonance point of the motor. A drastic reduction of torque occurs at motor resonance, and the stepping motor will stall if kept at this point for very long. (Some stepping motors have a fairly low resonance point.) Also, stepping motors must be accelerated slowly from a stopped condition or they will stall immediately. These limitations must be considered in proposed stepping-motor applications.

A TYPICAL STEPPING-MOTOR POSITIONAL SYSTEM

Figure 29-2 illustrates a block diagram of a typical stepping-motor positional system. The heart of the system is the preset counter. A number representing the desired position is loaded into this counter as the preset number. (If a computer is implemented as the controller, the computer will input this number.) The preset counter must have three basic digital outputs: A is greater than B (A > B), A is less than B (A < B), and A is equal to B (A = B). These three logic signals tell the stepper motor control to move forward, reverse, or stop.

The letter "A" stands for the *absolute position*. This is the actual position of the system at the present time. The letter "B" stands for the *desired position* for the system to reach. In other words, the "B" position is the most current preset number loaded into the preset counter.

The stepping-motor control is responsible for providing the proper sequential step pulses to accomplish forward and reverse rotation of the stepping motor. Like other motor controls, the step-

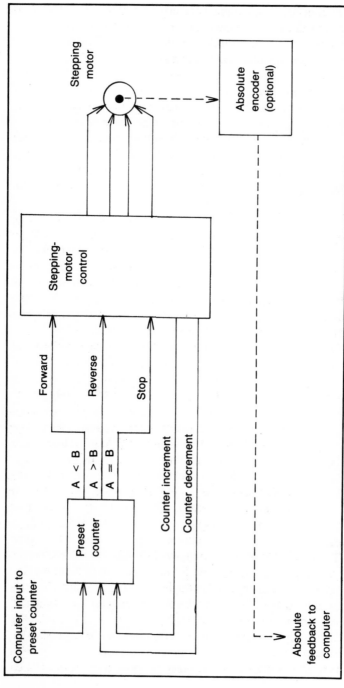

Fig. 29-2. Typical stepping motor positional system showing optional absolute encoder feedback.

ping motor control will probably incorporate some form of current-limit protection and maximum and minimum speed adjustments. It may also contain accelerate and decelerate adjustments.

The stepping-motor control must also output pulses back to the preset counter to increment or decrement the counter once for each step pulse provided. The "A" count will be incremented or decremented depending on the direction of movement.

Consider the step-by-step operation in Fig. 29-2. In this example, the stepping motor is gear-reduced so that each step pulse moves a workpiece exactly 1/1000 of an inch. Its present absolute "A" position is at 1.500 inches. The desired "B" position is also at 1.500 inches. Therefore, the A = B output is true, and the stepping motor is stopped. The computer decides to move the workpiece to a different position, so it inputs a new "B" number of 2.500 inches. (The decimal points in the "A" and "B" counts are for ease of understanding by the operator but are meaningless to the counters. Presently, in this example, the "B" counter is 1000 counts higher than the "A" counter.) The A = B output from the preset counter reads "not true." The A < B output reads "true," and the stepping-motor control begins to output step pulses, causing the stepping motor to rotate in the forward direction.

Each step pulse applied to the stepper motor causes the workpiece to move 1/1000 of an inch in the desired direction. Also, for each step pulse, a counter increment pulse is output from the stepper motor control to increment the "A" count in the preset counter by 1/1000 of an inch. In this way, the "A" count is continually updated to reflect the actual position of the workpiece. After 1000 step pulses have been applied to the stepping motor, it will have moved the workpiece to the 2.500-inch position. The "A" counter will have been incremented to 2.500 inches, which is the same as the "B" counter. The A < B output reads "not-true," and the A = B output reads "true" and stops any further stepping-motor movement.

If the number in the "B" counter happened to be smaller than the "A" count, the A > B output would read "true" causing the stepping-motor control to rotate the stepping motor in the reverse direction. For each step pulse applied to the stepping motor, a counter decrement pulse would be output to the preset counter to "decrement" the "A" count by 1/1000 of an inch. In this way, the position of the workpiece can be controlled and monitored in either direction.

This type of control system has one serious disadvantage. If a power failure occurs or the system power is turned off, the "A" count could be lost. Also, if a mechanical bind develops in the workpiece, causing the stepper motor to stall, the "A" count will become inaccurate, because the stepping-motor control does not have any positional feedback to tell if a motor stall occurred. For this reason, this type of stepping-motor positional control is referred to as *open loop*.

There are two commonly used methods of overcoming this problem. The obvious solution is to provide some form of positional feedback. Figure 29-2 shows an optional absolute encoder used for this purpose. Absolute encoders are not affected by power losses but are very expensive for the kind of accuracy demanded in many applications.

A more cost-effective solution to this problem is to provide a limit switch to represent a "home" position for the workpiece. If a power failure occurs, or the operator wants to check the positional accuracy, the operator simply presses the "home" push button. This causes the workpiece to begin to move toward the "home" (positional reference) position. When the workpiece physically hits the "home" limit switch, the "A" count in the preset counter is automatically reset to zero (or some preset number), and the correct relative position is established. The system will now maintain its positional accuracy until the stepping motor is stalled or the system power is lost.

VARIATIONS OF
STEPPING-MOTOR POSITIONAL SYSTEMS

For applications requiring high-speed stepping-motor operation, the system shown in Fig. 29-2 would overshoot the A = B signal. For this reason, high-speed systems require an additional decelerate-logic signal from the preset counter. Whenever the "A" and "B" counts are within some predetermined range of each other, the decelerate signal causes the stepping-motor control to decelerate to a speed slow enough to eliminate overshoot when the A = B signal goes true.

In some positional applications, incremental-type positional encoders may be used to sense motor stalls. These systems must still be homed in the event of a power failure, because incremental encoders cannot provide absolute positional feedback.

TROUBLESHOOTING
STEPPING-MOTOR CONTROL SYSTEMS

It is usually fairly easy to isolate the fault area in stepping-motor control systems. Obviously, the $A < B$, $A > B$, and $A = B$ logic signals must be present and accurate. If not, the preset counter should be suspected.

Most positional systems will contain emergency limit switches to disable the entire system if it tries to exceed certain physical limits. If the positioning system is setting at a physical extreme, simply homing the system may cure the problem.

If the correct logic signals are present to the stepping-motor control, but the stepping motor doesn't respond, there are several possibilities. Check all fuses and/or circuit breakers associated with the stepping-motor control. The stepping-motor shaft might be mechanically bound. This can be checked by removing the stepping motor to determine if it will run unloaded. If the stepping motor will not run unloaded, verify that the correct step pulses are being applied by the stepping-motor control. If they are, the problem is in the stepping motor itself.

In Fig. 29-1, the operation of a stepping motor is demonstrated using four switches (S1 through S4). In actuality, these switches are usually power transistors located in the stepping-motor control. It is common for one (or more) of these transistors to fail due to the high current requirements of most stepping motors.

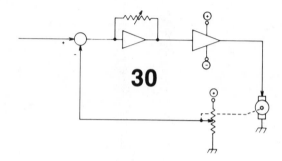

30

Robotics

THE CULMINATION OF ALL YOU HAVE STUDIED CAN BE DEM-
onstrated in the fascinating field of robotics. All of the sensors you have investigated can be used to familiarize a robot with
its environment. Continuous process-control techniques are utilized,
in conjunction with precision positioning systems, to give a robot
its mobility (if needed).

High-level CNC-type programs can be used as part of the logical mind of the robot. The same adaptive-control philosophies used
in some CNC equipment may allow a robot to adapt to changes in
its environment. In addition, other state-of-the-art techniques may
be employed to give a robot its "eyes" (imaging).

WHAT IS A ROBOT?

The Robot Institute of America has defined a robot as follows:

"A robot is a reprogrammable, multifunctional manipulator designed to move material, parts, tools, or specialized devices through
variable programmed motions for the performance of a variety of
tasks."

Consider the key points. To begin with, a robot must be
reprogrammable. In other words, it must have the capability of be-

Modern automated storage/retrieval is considered a form of robotics (courtesy of Litton Industrial Automation Systems Group).

ing programmed again if a new task is designated for it. This reprogrammability rules out using any kind of ROM program or hard-wired logical routine for the robotic brain due to the cost and inconvenience of changing the program. Therefore, a robot must have an extensive RAM memory for learning new tasks.

The term *multifunctional* indicates that the robot must be capable of performing more than one specific task. While performing these tasks, a robot must move (manipulate) objects such as tools or parts.

INDUSTRIAL ROBOTS

An industrial robot usually has a single arm to perform its designated tasks. The number of ways a robot can move is defined by how many axes of movement it has. For example, if a robot is capable of horizontal, vertical,and radial movement, it is said to have three axes of movement.

Structurally, there are only two types of robots: *polar* and *cylindrical*. In a polar-structured robot, the arm is capable of radial movement and possibly wrist rotation. The body pivots vertically or horizontally or both. In a cylindrical-structured robot, the body only rotates horizontally. The arm is jointed to facilitate the remaining axes of movement.

Industrial robots require sophisticated drive mechanisms to control their axes movement. Depending on the speed and payload (how much the arm can lift), these drive mechanisms may be *pneumatic, hydraulic,* or *electric.* Robots with electric drive mechanisms are the most expensive, but possess greater flexibility and payload. Pneumatic robots are the least expensive, but are limited in respect to payload. Hydraulic robots are the slowest of the three types.

High-technology robots incorporate extensive positional feedback. This gives the robot continuously variable positional control as well as programmable accelerate and decelerate functions.

The *end effector* is the robot hand which actually performs the desired task. If, for example, the robot is being used for a welding function, the end effector will be a welding gun. In many cases, the customer will custom-design the end effector for particular applications.

CONTROLLERS

For a robot to be intelligent (capable of learning and independent decision making), it must be controlled by an intelligent controller such as a computer or minicomputer (the controller may be a PLC in some cases).

There are two commonly used methods of teaching (programming) the controller. The first is the *walk-through method*. In this method, the operator places the robot in learn mode, then manually moves the robot arm through the movements to be performed. Meanwhile, the positional feedback from the robot records these movements. When all of the desired movements have been recorded (stored in the controller memory), the robot is capable of repeating these movements with high precision as often as required.

The second teaching method is the *software program.* Unfortunately, a high-level, universal robotic language has not been developed at the time of this writing. Commonly used high-level languages used for this purpose are FORTRAN, PASCAL, PL/1, and custom-designed languages from the robot manufacturers.

Appendix

NEMA Enclosure Definitions

WHEN SPECIFYING OR INSTALLING ENCLOSED INDUSTRIAL electrical equipment, the following information regarding NEMA (National Electrical Manufacturers Association) classifications will provide the user with industry standards. These descriptions are not intended to be complete representations of NEMA standards.

NEMA Type 1, General Purpose

To prevent accidental contact with enclosed apparatus; suitable for indoor applications where not exposed to unusual service conditions.

NEMA Type 2, Driptight

To prevent accidental contact, and in addition, to exclude falling moisture or dirt.

NEMA Type 3, Weatherproof (Weather Resistant)

Protection against specified weather hazards; suitable for use outdoors.

NEMA Type 3R, Raintight

Protects against entrance of water from heavy rain; suitable for general outdoor applications not requiring sleetproof material.